The Personal
Experience of Time

EMOTIONS, PERSONALITY, AND PSYCHOTHERAPY

Series Editors:
Carroll E. Izard • *University of Delaware, Newark, Delaware*

and

Jerome L. Singer • *Yale University, New Haven, Connecticut*

HUMAN EMOTIONS • *Carroll E. Izard*

THE PERSONAL EXPERIENCE OF TIME • *Bernard S. Gorman and*
Alden E. Wessman, eds.

A Continuation Order Plan is available for this series. A continuation order will bring delivery of each new volume immediately upon publication. Volumes are billed only upon actual shipment. For further information please contact the publisher.

The Personal Experience of Time

Edited by

Bernard S. Gorman
and Alden E. Wessman

Center for Research in Cognition and Affect and
The City College of the City University of New York

PLENUM PRESS · NEW YORK AND LONDON

Library of Congress Cataloging in Publication Data

Main entry under title:

The Personal experience of time.

(Emotions, personality, and psychotherapy)
"Most of the chapters in this volume had their origins as papers presented at the Eighth Annual Conference of the Center for Research in Cognition and Affect of the City University of New York. . . May 30, 1975."
Includes bibliographies and index.
1. Time — Psychological aspects — Congresses. 2. Cognition — Congresses. 3. Child psychology — Congresses. 4. Psychological research — Congresses. I. Gorman, Bernard S. II. Wessman, Alden E. III. New York (City). City University of New York Center for Research in Cognition and Affect.

BF468.P47	153.7'53	77-21964

ISBN 0-306-31039-2

© 1977 Plenum Press, New York
A Division of Plenum Publishing Corporation
227 West 17th Street, New York, N.Y. 10011

Printed in the United States of America

Contributors

Stuart Albert, Department of Psychology, University of Pennsylvania, Philadelphia, Pennsylvania.

Thomas J. Cottle, Children's Defense Fund of the Washington Research Project, Cambridge, Massachusetts.

Bernard S. Gorman, Center for Research in Cognition and Affect, City University of New York, New York. Present address: Department of Psychology, Nassau Community College, Garden City, New York.

William Jones, Department of Psychology, University of Pennsylvania, Philadelphia, Pennsylvania.

Robert Kastenbaum, Department of Psychology, University of Massachusetts—Boston, Boston, Massachusetts.

Klaus F. Riegel, Department of Psychology, University of Michigan, Ann Arbor, Michigan.

Arthur M. Schlesinger, Jr., The Graduate School and University Center, City University of New York, New York.

Gilbert Voyat, Department of Psychology, City College, City University of New York, New York.

Alden E. Wessman, Department of Psychology, City College, and Center for Research in Cognition and Affect, City University of New York, New York.

Preface

The fundamental nature of human time experience has concerned artists, poets, philosophers, and scientists throughout the ages. Any consideration of human action requires awareness of its temporal aspects. However, simply to view time in the same units and dimensions as the physicist employs in describing events robs personal time of its "lived" quality. The use of physical time concepts in the description of human events is often artificial and misleading. It fails to account for the facts that human time estimates rarely match clock and calendar time; that societies and individuals demonstrate vast differences in their constructions and uses of time; and that temporal perceptions and attitudes change within an individual both during a single day and throughout his life span.

The present volume does not view time as something that is sensed in the same way that one would sense or perceive spatial or sensory stimuli. Rather, it views time as a complex set of personally experienced cognitive constructs used by individuals and cultures to account for the order, the duration, and the organization of events. The authors in this book take a strong departure from earlier psychophysical studies of a "time sense" and address themselves to the uses and elaborations of time concepts in personal and social functioning.

The authors in this volume reflect a broad background of social-science approaches that, though diversified, seem to us to represent an underlying unity in their conceptions of time. All regard time as located in the interaction of social and natural contexts with the experiencing, aware individual who symbolically constructs meaningful representations of his world. Most view time as interpersonal as well as intrapersonal and consider temporal constructs as shaped by cultural and historical forces as well as by individual development. Most conceive of time as a phenomenological problem and feel that it is most usefully examined at the experiential level. Unlike many other approaches to time, the aim of these authors is not reductionistic and atomistic. Instead, the aim is to encompass the rich varieties and structures of temporality from a human perspective.

Most of the chapters in this volume had their origins as papers presented at the Eighth Annual Conference of the Center for Research in Cognition and Affect of the City University of New York and were presented at its meeting on May 30, 1975. We had been conducting our own research on the psychology of temporal experience and personality and were encouraged by our friend and colleague, the Director of the Research Center, John Antrobus, to organize and serve as co-chairmen of the conference on time. We were happy that almost all of our original choices for contributors to the meeting were willing and able to participate. They are individuals whose scholarly work on time had especially stimulated our thinking about the topic. Although not originally speakers at the conference, Stuart Albert, William Jones, and Arthur Schlesinger, Jr. kindly contributed valuable chapters to this volume.

We have received personal and intellectual support from many colleagues, particularly Donald Brand, Louis Gerstman, Elinor Mannucci, John Neulinger, David Ricks, Gertrude Schmeidler, Jerome L. Singer, Charles Smith, and Stephen Thayer. We thank Seymour Weingarten, Executive Editor of Plenum Press for his enthusiasm and encouragement.

Dale Gorman and Jane Wessman, our wives, have tolerantly put up with our trials and tribulations, our rantings and ravings throughout the course of this book and then helped to place our lives in meaningful and rewarding temporal perspectives.

BERNARD S. GORMAN
ALDEN E. WESSMAN

New York City

Contents

- *Chapter 2*

Toward a Dialectical Interpretation of Time and Change
Klaus F. Riegel

Chapter 3

The Temporal Transition from Being Together to Being Alone: The Significance and Structure of Children's Bedtime Stories
Stuart Albert and William Jones

Chapter 4

Perception and Concept of Time: A Developmental Perspective
Gilbert Voyat

Chapter 5

The Time of Youth
Thomas J. Cottle

Chapter 6

**Memories of Tomorrow: On the Interpenetrations of Time
in Later Life**
Robert Kastenbaum

Chapter 7

Images, Values, and Concepts of Time in Psychological Research
Bernard S. Gorman and Alden E. Wessman

Chapter 8

The Modern Consciousness and the Winged Chariot
Arthur M. Schlesinger, Jr.

1

Editors' Introduction

This chapter presents the basic perspectives necessary to an understanding of human temporal experience and concepts. It focuses on the psychological nature of temporal experience and discusses the emergence of temporal capacities in evolution and in human cognitive and social development. It first considers the phylogenetic expansion of temporal capacities, focusing on man's distinctive biosocial characteristics and cognitive abilities that are basic to the emergence of symbolic and relational thought. It emphasizes the role of imagery and symbolic representation in remembering, anticipation, and planning. The genesis of temporal awareness is next considered from the perspectives of genetic epistemology, psychoanalytic ego psychology, linguistics, and phenomenology. While temporal awareness may seem directly apprehended and intrinsic in experience, it is more appropriately viewed as a complex of developed abstract-conceptual frameworks gradually constructed and acquired. To understand the emergent character of time as a set of cognitive construct systems, we must recognize the intimate relation of temporal constructions and awareness to personal identity and to sociocultural and historical awareness. This first chapter presents a framework for relating the contributions of the subsequent chapters and placing them in perspective.

The Emergence of Human Awareness and Concepts of Time

Alden E. Wessman and Bernard S. Gorman

We all know intimately the passage of time. Life is transitory: hours fly by, days merge into years. Our personal experience seems always located in time. "Now" possesses an immediate presence and reality that past and future lack. But memories endure, often repossessing us with vivid detail and feeling. And hopes and fears can transform the present with images of beckoning paradises or ominous hells. While ever in the present, our awareness and imagination transcend it. Yet we know we are not the masters of time; rather it restricts and coerces us. Though awareness and conceptions of time are products of the human mind, time itself seems to possess an existence apart, its passage impersonal and inexorable. As an old Italian proverb put it, "Man measures time and time measures man." This intimate and personal, yet aloof and detached character constitutes the paradox of human time.

Thus, the nature of time and temporal experience has remained a persistent puzzle, challenging thinkers throughout intellectual history. Repeatedly the question has been posed: What is time and how do we come to know it? Notable contributors to this continuing inquiry include Aristotle, St. Augustine, Newton, Kant, Guyau, Bergson, Janet,

Alden E. Wessman • Department of Psychology, City College, and Center for Research in Cognition and Affect, City University of New York, New York. **Bernard S. Gorman** • Center for Research in Cognition and Affect, City University of New York, New York. Present address: Department of Psychology, Nassau Community College, Garden City, New York.

James, Freud, Einstein, Husserl, Heidegger, Minkowski, Merleau-Ponty, Cassirer, and Piaget. Fortunately, the terrain has already been well surveyed by some outstanding scholars (Doob, 1971; Fraisse, 1963; Fraser, 1966, 1975; Whitrow, 1963, 1975). They have reviewed much of the vast philosophical and scientific literature on time and clarified many important aspects.

Here we attempt to formulate, as much for our bemused selves as for the reader, some basic psychological perspectives that emerge as fundamental to an understanding of human temporal experience and concepts. We are not so bold as to try to grasp the cosmic nature of time as a basic feature of the universe intrinsic in phenomena at all levels. Rather we are "simply" attempting to understand the psychological nature of human temporal experience and to consider how concepts of time arise in personal and social development.

1. Evolution and the Expansion of Temporal Capacities

All forms of life seem to exhibit some degree of temporal coordination. Rhythm, periodicity, and adaptation to changing conditions are inherent in organic life at every level. Highly complex temporal capacities have successively emerged in the course of evolution. Living organisms manifested increasingly elaborate adaptive aims in an expanding encounter with wider and richer environments. Weston La Barre (1954) and Julius T. Fraser (1975) have argued that temporal capacities should be viewed in the context of evolutionary developments that show a progressive expansion of purposeful activities. Animals, with their capacity for autonomous movement, actively engage their surroundings. Locomotion in the service of new modes of adaptation is related to purposive exploration and environmental encounter. In evolutionary perspective, more active and differentiated goal-seeking and danger-avoidance appears linked to the greater adaptive "intelligence," with greater awareness of the surrounding world and increased capacity for appropriate response, displayed by higher forms of animals.

Vertebrates, particularly mammals, possess acute awareness of their environments based on fundamental improvements in both sensory capacities and effective neuromuscular response. Advances in the structure and function of the central nervous system and the brain coordinate complex sequences of environmental search and response. Evolution shows fairly consistent progression toward more varied and sensitive receptors, which register adaptively significant events at a distance, and more complex brains and effectors, which can coordinate increasingly effective patterns of action. Higher animals actively attend to and respond to their environments and possess greatly developed capacities to learn, remember, and anticipate. This increased responsiveness and more effective coping is necessarily accompanied by an enormous extension of temporal capacities. Undoubtedly, increased capacity for inner representation and mental imagery that can hold, maintain, and provisionally envision currently nonexistent states is of critical significance in this advance.

2. Human Biosocial Distinctiveness and Symbolic Thought

In the primate evolutionary sequence of lemur–monkey–ape–man, as Weston La Barre (1954, pp. 50–54) has vividly described, offspring progressively tend to be smaller, more helpless, and more immature at birth. The period of infantile dependency lengthens, social independence and sexual maturity arrive later, and the significance of postnatal learning is markedly increased. The human animal is a distinctive creature with unusual sensory and motor abilities and, most important, an enormously complex brain with vast capacity for varied learning and symbolic representation and manipulation. The sociable, babbling human infant is unique in his potential capacity for speech and communication through the acquisition of language and symbols.

The distinctively human evolutionary advantages permit the learning of language and the development of communication by means of socially shared symbols and abstract relational schemata. The capacity to utilize symbols and gradually develop and acquire elaborate systems of cognitive constructs appears generic in and intrinsic to the human mind. While other primates have been found to have some limited capacity for symbolic thinking, the extent and range of such abilities and their capacity to serve as a means of communication appear uniquely human. The manipulation of symbols and mental constructs allows internal representation of, manipulation of, and experimentation with "external reality." The capacity for intentional action and future planning seems to require such representational schemata and their maintenance and manipulation.

External reality is not uniform for and directly known to all species but rather appears to be "experienced" through their perceptual and response potentialities and capacities for assigning significance and meaning to environmental events. These capabilities vary widely among different organisms. Jacob Von Uexküll postulated that each species has a "world" of its own and an "experience" of its own based on the potentialities of its receptor and effector systems. As both Schachtel (1959, pp. 69–70) and Fraser (1975, p. 75) have argued, this hypothesis implies that there are many levels and varieties of temporalities experienced by creatures on different evolutionary levels possessing different sensory capacities. It is necessary to recognize and distinguish various perceptual universes and temporalities and to acknowledge the active contribution of the perceiver to the nature of the world as known to him.

Ernst Cassirer (1944) and Susanne Langer (1951) have argued that human beings are unique in the complex functioning of elaborate symbolic systems. These are mediated by the operations of the greatly expanded central nervous system interposed between the receptor and effector systems found in all higher animals. Compared with other animals, humans live not merely in a broader "reality" but in a new and transformed dimension of "reality." Rather than fairly direct, automatic, and immediate responses to external stimuli, there are delay and interruption through central cognitive processing involving complex and flexible systems of symbolic representation and meaning. Humans do not exist simply in a physical world; they experience and inhabit a symbolically transformed universe. The implication of these ideas is that there is not a single objective reality and a uniform, universal temporality. Rather, there are different levels and forms of temporal relation and experienced temporality in nature and organic life.

Our own view of human time regards it as an emergent set of cognitive construct systems based on personal experience and identity, influenced greatly by sociocultural and historical awareness. While temporal awareness may seem directly apprehended and intrinsic in experience, it is more appropriately viewed as features of complex abstract conceptual frameworks gradually constructed and acquired.

3. Images, Symbols, and Relational Thinking

Images, symbols, and language are the cognitive tools used to construct the elaborate and distinctive human worlds of thought and culture. Without a complex system of images and symbols, relational thought cannot arise at all, much less reach full development. Reflective thought arises from man's capacity to single out from the continuous stream of sensory phenomena certain "fixed" elements or relations in order to isolate them and to concentrate attention upon them. This capacity for abstract relational thought is only gradually realized in human development. The increasing complexity of the nervous system and cerebral organization, particularly in cortical association areas, appears to underlie the striking evolutionary shift from primarily reflexive to deliberate, purposive behavior. Remembering and anticipating are possible only when mediating mechanisms in the brain enable the organism to represent or imagine objects or events that are not present in its immediate sensory environment. While other higher animals show some limited capacities for remembering, anticipating, and planning, humans appear unique in the extent of such abilities. An essential feature in this advance is symbolic representation by means of imagery, speech, and the elaborate cognitive schemata and constructs gradually acquired through experience. As Cottle and Kleinberg (1974, pp. 3–35) have pointed out, while all other animals live primarily in the present, it is extremely difficult for adult human beings to immerse themselves completely in the present for any prolonged period. Associated events from both past and present continually intrude into consciousness and are integrated into ongoing patterns of action and meaning.

4. The Genesis of Temporal Awareness in Human Action

With this appreciation of the distinctive features and the fundamental significance of human symbolic capacities as essential for relational thinking, let us turn more directly to consideration of how temporal awareness and concepts of time emerge in human development. In the course of the long history of speculation concerning the nature of time (Fraser, 1975; Whitrow, 1963), many thinkers have argued that despite its presumed intimate association with and conformity to the "reality" of a postulated universal order, ultimately the idea of "time" must be acknowledged as having its origin in the mind of man. St. Augustine, the great pioneer of the study of internal time, concluded that we can measure time only if the mind has the power of holding within itself the impressions made on it by passing events. The British empiricists, considering the origins of time, agreed that it was the succession of ideas in the mind. Immanuel Kant, though seeing time as an *a priori* category not derived from experience, also emphasized its location in the observer. It is clear that the consensus of most modern thinking in philosophy and the sciences has been to question the absolute time conceived by Newton and rather to view time in a relativistic framework inevitably tied to the observing subject.

Since Jean-Marie Guyau's *La Genèse de l'Idée de Temps* in 1890, and particularly with Jean Piaget's profound contributions to genetic epistemology, psychologists have come to regard time not as a given *a priori* category of the human mind. Rather, it is viewed as a set of gradually emerging cognitive constructs and relational schemata derived from experience and action. In Guyau, one finds an early statement of the ideas that Piaget has so convincingly established and extended, namely, that the origins of human temporal concepts arise in the infant's sensorimotor "experiments." Active seeking and envisioning of goals in the environment are associated with the sense of purpose and the associated effort that is the ultimate source of ideas of cause and effect. In Guyau's view, every need or desire implies the possibility of satisfaction, and the aggregate of these envisioned possibilities leads to the concept of the future (Whitrow, 1963, pp. 50–53). As Guyau (1902) put it: "We must desire, we must want, we must stretch out our hands and walk to create the future. *The future* is not *what is coming to us* but *what we are going to*" (p. 33, quoted in Fraisse, 1963, p. 173).

Temporal concepts are thus seen as mental constructs derived from experience with the consequences of purposive action in the environment. Human ideas of space and time are emerging conceptions that can be understood only in the perspective of their development and their relation to purposive action. In the view of most contemporary psychologists, temporal awareness and conceptions of time are products of human evolution and individual cognitive development. While awareness of temporal phenomena may seem directly inherent in primary and irreducible personal experience, it is more appropriately viewed as an abstract conceptual framework that is gradually acquired and constructed by each individual. This conception of time is shared by the contributors to this volume and the epistemological significance of this view will be considered in depth in the chapters by Klaus F. Riegel and Gilbert Voyat.

In considering the development of temporal awareness and concepts of time, we review here relevant ideas from a number of the main schools and approaches in modern psychology and social science. They include contributions from psychoanalysis and psychoanalytic ego psychology, genetic epistemology, language development, and cultural and historical studies. Although it is difficult to synthesize and unify all these diverse contributions, we present what in our view appears most basic and meaningful.

5. Psychoanalytic Conceptions of Ego Development and Temporal Awareness in Infancy and Early Childhood

In the opinion of some of the most thorough and informed scholars on the psychology of time (Doob, 1971, pp. 297, 319–320; Wallace & Rabin, 1960, pp. 214–215), the psychoanalytic account of the genesis of temporal awareness is not consistently formulated and cannot be considered a well-validated body of knowledge. Certainly, there is a beguiling clinical literature vividly illustrated with patients' bizarre temporal attitudes and experiences. And there has been ample theoretical speculation variously relating the origins of temporal attitudes to particular features of psychosexual stages and socialization. While favorably inclined to this school, we do agree with the critics that psychoanalytic accounts of time leave much to be desired. The statements of various psychoanalytic writers are quite varied and are admittedly difficult to validate. Nonetheless, despite these obvious problems, psychoanalytic conceptions are definitely of value in our opinion.

Clearly, Sigmund Freud was the first major psychological theorist with an elaborated conception of early mental development that has profound implications for an understanding of the origins of temporal awareness. In his theories, Freud postulated an initially "timeless" unconscious id and regarded temporal awareness as having its origin in the gradual development of conscious ego functions.

The main features of Freud's theories relevant to time were his early characterization of primary and secondary processes and his later conception of the gradual development of the ego as an organized executive agency operating according to the secondary process (Brenner, 1974, pp. 45–51; Fenichel, 1945, pp. 15–18, 33–53). The primary process was so designated because Freud held that it was the initial disorganized and irrational mode of mental functioning. He postulated that throughout life, the id continues to operate according to the primary processes of the unconscious. Though fundamental in mental life, Freud considered the id an obscure and inaccessible region of the personality revealed mainly through the psychoanalytic study of dreams and neurotic symptoms (Freud, 1933, pp. 103–105). He described the id's mode of operation as chaotic, disorganized, impulsive, contradictory, and unfettered by the constraints of reality. He wrote that in the id, there is nothing corresponding to the idea of time and no recognition of its passage. He held that repressed impulses and experiences are virtually immortal and are preserved for years as though they had only recently occurred.

The secondary processes of conscious rational thought develop slowly, but progressively, during the early years of life and come to characterize the reality-oriented operations of the mature ego. The ego utilizes the sensory apparatus of the "perceptual conscious system" to register significant events, both external and internal. It gradually comes to construct and maintain a reasonably accurate representation and assessment of the world. Freud (1950) wrote that while unconscious mental processes are "timeless," the abstract conception of time seems to be wholly derived from the mode of functioning of the "perceptual-consciousness system." Besides its sensory and perceptual features, "the ego controls the path of access to mobility, but it interpolates between desire and action the procrastinating factor of thought, during which it makes use of the residues of experience stored up in memory" (Freud, 1933, p. 106).

In Freudian theory, the investments of psychic energy, or cathexes, attached to mental representations of objects in the form of images, memories, thoughts, and fantasies are initially very weak and labile (Brenner, 1974, p. 18). The perpetually immature id, operating by the primary process, seeks immediate gratification and can readily shift from its original object or method of discharge and find substitute gratifications. The gradual establishment of the ego, operating according to the secondary process, establishes a different temporal order. The emphasis is more on the ability or capacity to delay the discharge of cathetic energy until environmental circumstances are favorable for gratification. Also, the emotional cathexes become much more firmly attached to particular objects or methods of discharge than was the case with the primary process.

Freud was quite cautious and provisional in the few places where he directly stated his ideas concerning the genesis of temporal awareness. Evidently, he held that the tentative establishments of and fluctuations of object attachments and cathexes in early infantile development were "at the bottom of the origin of the concept of time" (Freud, 1953b, p. 180). The perceptual-conscious system, the foundation of the ego, was seen as the apparatus through which impulses passed to cathected external objects and stimulation from the outside world returned to what Freud termed "unconscious memory systems." The formation and disruption of such early relations to gratifying objects in the external world would thus appear to lie at the basis of rudimentary awareness of temporal relationships.

Freud's emphasis on early object attachments was followed by many subsequent analysts (Wallace & Rabin, 1960, p. 214; Yates, 1938), who attributed developing awareness of time to the patterns of bodily rhythm that become established during the oral period. These rhythmic patterns become related to the periodic frustrations and satisfactions of bodily needs (particularly the intake of food) and provide the basis for the initial development of the sense of time.

The early psychoanalytic formulations were, of course, largely hypothetical reconstructions of probable stages of infantile development. Work more closely based on direct observations of infant development is perhaps best exemplified by René Spitz (1963), who was particularly concerned with early emotional development in the infant. Spitz emphasized the significance of object attachments in the initial stages of ego development and the emergence of consciousness. The ego, which slowly develops as the central coordinating organization, enables the infant eventually to recognize his own emotions and to relate their expression to instigating stimuli and situations. Spitz held that such organization begins in a rudimentary fashion around the third month and, for a considerable period, is only loosely linked to specific types of experience. By the third month, the innate "turning-toward" response initially manifest in the sucking reflex has become linked to the infant's anticipation of pleasure from a gratifying adult and is expressed by the smile directed toward the adult face.

According to Spitz's observations, once the smiling response is reliably established, it ceases to be random and becomes stimulus-specific. Beginning around the third month of life, smiling consistently occurs in situations of need gratification. Spitz held that when the infant has established the smiling response, he has also achieved conscious perception and becomes capable of rudimentary mental representation. Smiling becomes a specific, nonrandom response to specific stimulus situations.

By the fourth month, Spitz observed, the infant cries when an adult leaves after playing with him, just as he reacts by smiling when the adult approaches him. Thus, besides the unconditioned stimuli of pain and discomfort that originally gave rise to distress and crying, the infant has now established a "quasi need" for specific-object contact with a gratifying adult and "proleptic (or anticipatory) function." Once the function of anticipating pleasure is acquired, it becomes possible for the child also to anticipate unpleasure when losing the gratifying object.

Thus, according to Spitz, between the ages of 3 and 6 months, the anticipation of unpleasure has acquired a signal function, like that described by Freud in his second theory of anxiety (Brenner, 1974, pp. 69–80). This overt emotional expression of mounting distress is experienced as a danger signal by the incipient ego of the infant. In addition, the affect functions as an interpersonal communication of distress—though at first unintentionally. These functions of internal signal and external communication become increasingly effective in triggering adaptive measures by the ego and in communicating with the surround. These signaling processes eventually lead to the establishment of specific emotions firmly linked with their appropriate expression. Thus, Spitz's account of early ego development emphasized the signal function and the anticipatory features of affect in self-communication and social communication about the status of important interpersonal object relationships. The infant gradually establishes continuing emotional ties to the significant people in his world and becomes increasingly able to maintain and influence those important ties.

Other important theorists of the psychoanalytic ego-psychology school have similarly emphasized the importance of early object relations and the critical role of affect in the development of active anticipation and reality-oriented thought (Hartmann, 1964, pp. 114–115, 245–247). While a great many observers of child development regard temporal awareness as emerging out of early emotional ties to objects, we must recognize that the exact sequence and character of these stages is only beginning to be adequately conceptualized and that much remains to be understood and carefully formulated (Bowlby, 1969, pp. 265–330).

The clinical literature of psychoanalysis frequently attributes various difficulties in temporal orientation and attitude to a patient's fixations and conflicts stemming from the oral stage. The formulation of such connections began with Sigmund Freud and Karl Abraham's interpretations relating adult character types and orientations to the psychosexual stages (Abraham, 1927, pp. 370–476). Probably the most frequently cited temporal characteristic attributed to the oral stage is the impatience of narcissistic individuals, who are highly dependent upon external sources of gratification for security and self-esteem and who find it very difficult to tolerate delays of gratification or frustration of their wishes. This kind of "oral character" is regarded as typical of alcoholics and other addicts, impulsive personalities, and depressives (Fenichel, 1945, pp. 367–390). The desperate craving and hopelessness experienced by these and other types of patients is an aspect of "basic mistrust," as described by Erik Erikson (1963, pp. 247–251).

On the other hand, the security and confidence coming from gratifying experiences in infancy establish a lasting sense of "basic trust," which has important temporal features. In Erikson's view, the infant's first social achievement is his willingness to let the mother out of sight without undue anxiety or rage, because her presence has become an inner certainty as well as an outer predictability. He holds that such consistency, continuity, and sameness of experience help establish a rudimentary sense of ego identity that depends on an inner population of remembered and anticipated sensations and images firmly correlated with the outer population of familiar and predictable things and people. Trust also implies not only that the child has learned to rely on the sameness and continuity of the outer providers, but also that he can trust himself and the capacity of his own organs to cope with urges.

As we have seen, Freud and many subsequent psychoanalytic theorists placed the origins of temporal awareness and attitudes in early ego development during the oral stage. But Freud and other analysts also discussed temporal characteristics developed in later psychosexual phases, particularly the anal stage. These additional temporal attitudes are seen as arising from a further development of the coping patterns necessary for dealing with postponed gratification. In the second year of life, with increasing sphincter control, both the expulsion and the retention of feces are regarded as sources of erotic pleasure and satisfaction to the child and of potential conflict with parental attempts at toilet training. The resulting feelings may become generalized ways of relating to important external objects. Thus, some of the symbolic meanings of anal expulsion include giving or losing or rejecting and defiling, while retention may symbolize possessing or controlling. Freud (1953a) and Abraham (1927) postulated two main character types as developing from conflicts and fixations of this period. The "anal expulsive character" was held to be defiant, cruel, wantonly destructive, messy, and disorderly. The "anal retentive character" was seen as meticulous, stingy, greedy, orderly, prudish, and obstinate.

Psychoanalytic descriptions of neurotic "anal personalities" picture them as disturbed in their attitudes toward time (Fenichel, 1945, pp. 282–283). They may be stingy or prodigal with their time, or both alternately. They may be excessively punctual or unpunctual. The obsessive-compulsive is seen as typically manifesting great conflicts and ambivalence with respect to scheduling and employing time. The anal period is seen as a critical stage in which the child acquires ideas of order and disorder regarding time and of the measurement of time in general. While the oral stage is seen as fundamental in establishing initial temporal awareness and anticipation, anal experiences are seen as important in developing schedules as a means of mastering reality. Neurotic disturbances in the practical use of time are seen as typical of compulsion neuroses with unconscious anal conflicts.

Although most psychoanalytic theorizing about time focuses on the oral and anal stages, it would be incorrect to state that later stages are entirely neglected. Some psychoanalytic writers have discussed developments in later stages. Edith Jacobson (1964, pp. 119–137) particularly emphasized the significant role of superego development during the Oedipal and latency periods in establishing lasting identifications, in regulating drives and defenses, in implanting long-term goals and ideals, and in generally reenforcing the sense of time. And as we discuss later, Erik Erikson has much to say concerning temporal perspectives in adolescence and later life. But first, we wish to consider other important contributions to our understanding of the development of temporal awareness in infancy and childhood.

6. Piaget's Observations on the Temporal Aspects of the Sensorimotor Stage

Jean Piaget's investigations and theory of cognitive development have greatly advanced the understanding of all aspects of human thought, including temporal experience. He has discerned four major stages with their approximate ages: the *sensorimotor* (birth to 2 years), the *preoperational* (2–7), the *concrete operational* (7–11), and the *formal operational* (12 through adulthood). For each stage, Piaget has provided detailed descriptions of the characteristics of children's cognitive activities, the processes that appear to underlie them, and their implications for children's understanding and conceptions of the world (Elkind, 1967, 1974; Langer, 1969, pp. 111–148; Phillips, 1975, pp. 23–60; Piaget, 1954).

Piaget's view of cognitive development is as an integrated process involving maturation of the neurological system, interaction with the physical world, and social communication. His account is neither "environmentalist," in which learning and experience mold a passive organism, nor "nativist," in which intellectual capacities exist preformed and simply unfold in the course of development. Rather, Piaget has demonstrated how even the most basic features of perception and thought are the result of constructive activities of the child in interaction with sensory data and experience. Knowledge and understanding are mental constructions that gradually emerge from the child's repeated active encounters with the world.

Piaget's view of cognitive growth holds that motor activity is the initial basis for mental operations. In broad outline, the sensorimotor period can be characterized as showing the establishment of a functional knowledge of ends-and-means relationships and the beginnings of a rudimentary understanding of space, time, and causality. A major accomplishment of infancy is the recognition of and memory for objects. In the early months of life, the infant gradually comes to attend and respond to objects, and then he actively seeks out and interacts with them. Initially, the infant deals with objects as if their existence were dependent upon their being present in his immediate perception—"out of sight, out of mind." Toward the end of the first year, however, the infant begins to seek objects even when they are hidden. This accomplishment appears to require some continuing internal mental representation of the absent object and seems to be one of the clear early manifestations of mental imagery and symbolism critical to later cognitive growth.

We wish to examine in some detail Piaget's rich account of the sensorimotor developments that underlie the complex transition from the relatively helpless infant—faced with a bewildering and chaotic collage of undifferentiated, impermanent, unrelated environmental

events—to the actively questing, semiverbal, and socially responsive toddler who has become fairly competent in managing his world. At birth,
except for periodic physiological tensions and a few simple reflexes,
there is little evidence of organized or directed cognitive activity. During
the next few months, "circular" or repeated action patterns gradually
develop, and there are differentiated responses to various objects
(Phillips, 1975, pp. 27–29). Reaching and touching are repeated time
and again with evident delight. During this period, the infant's activity
begins to be centered on objects, but there presumably is no representation of "objective" reality—no generalized space or time, no permanence of objects. There would seem to be only fleeting events that are
transient aspects of the infant's activity. When an object in his field of
vision disappears, it ceases to exist for him. Without permanence of
objects, there presumably is no generalized space—though there may be
the vague beginnings of visual space, auditory space, and kinesthetic
space. Temporal dimensions would appear to be rudimentary: with little memory of objects, there can be little recall of the past; with very
simple goals and limited sensorimotor schemata, there can be little
anticipation of the future.

Toward the middle of the first year (approximately 4–8 months),
limited intentions and means–ends relationships begin to become more
manifest (Phillips, 1975, pp. 29–34). The infant discovers that certain
actions produce gratifying consequences, and he attempts to repeat
them. There is increasing coordination of seeing, reaching–grasping,
and sucking. Objects begin to have a permanence beyond immediate
perception—when an object that he has been attending to is suddenly
removed, the baby may briefly search for it, rather than simply shifting
his attention to something else, as he formerly did. But the search for
the absent object is of very brief duration and is confined to one
modality; for example, a felt object is groped for but not looked for, a
seen object is looked for but not groped for.

The end of the first year (8–12 months) shows a much clearer
emergence of intentionality and effective means–ends coordination
(Phillips, 1975, pp. 34–39). There is more persistent determination to
reach attractive objects despite the interposition of simple barriers to
action. Instead of losing interest or being distressed if an obstacle is
placed between him and a desired object, the baby may now displace
and remove the obstacle. Intentionality of action and separation of
means from ends can be more definitely inferred from the child's

behavior. The search for hidden objects becomes more protracted. Presumably, then, past memories and future anticipations are becoming more steadily maintained and more coordinated with effective action patterns. There appears to be some appreciation of causality by outside agents, as the child sometimes signals and waits for adults to do things for him.

In the first half of the second year (12–18 months), according to Piaget, the child becomes much more experimental in his attempts to discover new properties of objects and events (Phillips, 1975, pp. 39–45). He engages in a pattern, already very familiar to his parents, of letting go of objects or throwing them in order subsequently to pick them up; however, the aim now is not mere motor activity but observation of the various results. Instead of simply activating the motor pattern in a stereotyped manner, now the child deliberately acts upon the environment to find out what happens and varies his actions to produce different effects. Intentionality and means–ends relations have thus become quite advanced in the realm of bodily actions. Also, a new level of object permanence is attained with correspondingly altered perceptions of spatial relationships and temporal sequences. Now the child can correctly rediscover hidden objects in various locations as long as their placement has been directly observed by him (although he still has trouble with sequential displacements if the object is not directly observed). There clearly appears to be internal representation of objects with a unity and independence of situations not previously achieved. In the solving of simple motor problems, there is gradually more effective discrimination of appropriate actions and fewer random or misdirected trial-and-error approaches. Successful solutions to similar situations are fairly readily utilized, though they may not be very successfully generalized to more dissimilar ones. It is clear that there has been a shift from stereotyped behavior to a kind of systematic variation of responses.

The last substage of the sensorimotor period (approximately 18–24 months) shows the invention of new means through mental combination (Phillips, 1975, pp. 46–51). Whereas earlier stages showed the application of familiar schemata to familiar and then new situations, and then modification of familiar schemata to new situations, this stage shows the invention of new means of solving problems through the reciprocal assimilation of schemata. The child, like Wolfgang Köhler's (1959) chimpanzees, who learned to use sticks to bring desirable objects within

their reach, shows "insightful" solutions—but they reflect a new relationship among previously developed cognitive schemata built up
through countless interactions of the child with its environment. Now
much of the experimentation is interiorized, with little overt motor trial-
and-error but presumably with inner imaginal representation before
effective action. Object permanence is now maintained through successive displacements, even when the object is out of sight. The child
may effectively retrieve a ball by going around to the other side when it
has "disappeared" by rolling under a piece of furniture. Such abilities
indicate the increasing importance of internalized symbols in the child's
construction of space. The construction of temporal relations, though
not yet abstractly formulated and verbalized, have undergone transformations whereby internal symbols make possible both memory of
past events and anticipation of future ones. Understanding and effective
utilization of cause–effect sequences are much advanced and are
dependent upon the extension of time and space coordinates. The child
has come to envisage and execute complicated sequences of motor
actions extending over a long span of time. In such a fashion, the temporal perspectives appear to originate and develop during the
sensorimotor period.

It is evident that there are many important basic parallels between
the psychoanalytic and Piagetian accounts of the early development of
temporal awareness. Both conceive of the infant as initially living in an
objectless and timeless world, with only recurrent physiological tensions, unorganized sensory capacities, and fairly rudimentary motor
patterns. Awareness of time is postulated to develop slowly and concur-
rently with the gradual establishment of coordinated motor activities
from repeated goal-oriented encounters with the world. Slowly, the
character of physical and human objects is established and maintained
through persisting mental imagery and symbolic representation. The
building of these reality-oriented representations of the object world and
the coordinated spatial and temporal schemata in which they are
embedded is the major cognitive accomplishment of early childhood.

7. Language, Symbols, and the Development of Temporal Schemata in Childhood

Thus far, we have mainly considered the emergence of temporal
awareness in the preverbal child. The acquisition of and the develop-

ment of facility with language and symbols, the distinctive human attainment, constitute a major advance in the child's ability to understand and communicate. The character of language and symbolic systems, their role in thinking and their manner of acquisition, are challenging topics with fundamental significance. We cannot here adequately review the enormous literature on language (Black, 1968; Bram, 1955; Carroll, 1964; Church, 1966) and symbolism (Cassirer, 1944; Langer, 1951; Pollio, 1974; Royce, 1965). But we will try to indicate their critical role in the child's developing capacity to master conceptual relationships and to acquire the very abstract notion of time as a general frame for all phenomena and events.

Language has been defined as "a structured system of arbitrary vocal symbols by means of which members of a social group interact" (Bram, 1955, p. 2). It serves as a symbol system whereby individuals can converse with others and themselves. By means of language, people convey information, thoughts, and feelings to one another and influence each other's actions. They also employ it in their own internal communication or "thinking." The individual can respond to his own inner speech and ideas: he can, for example, respond to verbal representations of previous or anticipated situations and relations, consider their implications, and act accordingly. Many words represent concepts standing for abstract aspects or features of mental operations that have, in varying degree, practical application to specific actions and situations in the "real world."

Through spoken and, later, written language, the scope of human communication becomes enormous. Korzybski (1933) wrote of the "time-binding" function of language in the transmission of ideas, beliefs, techniques, and attitudes from one generation to another. For good or ill, the thoughts of the past remain with us and shape our actions, individual and collective. Language is an important vehicle for representing and communicating about past, present, and future. Not all representations of time are necessarily linguistic, but language is a very advanced and efficient means for expressing temporal relationships. Words render us susceptible to the range of human understandings and illusions, hopes, and fears. Humans do not live merely in the sensorimotor world of relatively simple activities and events, they live with and through the extended awareness and the enduring significance created by symbols and language.

The child learns and discovers the characteristics of "reality" as he develops. Language, including both what other people tell him and what he comes to tell himself, plays a critical role in this discovery and his apperceptions and conceptions of the world as coherent and meaningful. The world perceived and understood by adults includes not only objects and space defined in terms of basic sensorimotor activities, but elaborate and integrated conceptions of complex aspects relating objects to space, to each other, to past and future conditions, and to their personal significance. A major difference between infant and adult experience is in the awareness and conceptualization of the relational structures of reality (Church, 1966, pp. 14–15). Temporal relationships and sequences are not necessarily clearly perceived or understood initially. Many complex cause–effect relationships obviously cannot be readily understood by the child, as when there is a long temporal delay between cause and effect (e.g., tending a plant and its eventual blooming) or when causal agents or resultant conditions are not obvious (e.g., the striking of a clock or immunity following an innoculation), and interpersonal transactions may be highly obscure. But throughout childhood, the individual builds and acquires the abstract relational schemata for understanding and action, much of it mediated by words and verbal symbols concerning time.

The notion of *schemata* was introduced by the neurologist Henry Head as a neurological construct and figures prominently in the theories of Bartlett, Piaget, and other cognitively oriented psychologists (Church, 1966, pp. 36–37). A schema is a mental construct or principle employed to organize experience. Schemata are cognitive constructs that formulate the regularities in the way things are constituted and behave, so that we come to perceive the environment as coherent and orderly. Schemata "exist" in the organized patterns of thinking and action that allow us to cope with the world in terms of its apperceived characteristics and meaning. Integrated and organized activity requires orientation in terms of such spatial, temporal, and situational frameworks. Schemata maintain the stability of our environment and our location and potentialities for effective action within it.

In reviewing Freud's and Piaget's work, we attempted to describe some of the features in the initial schematizations of temporal awareness established during infancy. We now wish to see how the acquisition of language builds upon and extends these beginnings of temporal conceptualization. The catalog of developmental achievements is impressive. During the childhood and school years, the individual acquires an orientation toward time that comes to include: the abstraction of time from space and activity patterns, a massive vocabulary of words with temporal reference, an understanding of the scales used to measure time's passage, a conception of growing up and growing old, knowledge of the historical past and future possibilities, and the ability to order and coordinate his activities in purposeful and effective sequences.

The child's early attempts at active communication in the second half of the first year are quasi-symbolic, and he devises ways to tell people what he wants by a process of concrete enactment: to be picked up, he holds out his arms to an adult; to get an adult to operate a toy, he hands it to him. With respect to genuine symbolic language, passive understanding precedes active speech. The child begins to respond appropriately to cue words connected with familiar games like "peek-a-boo," with routine activities like "bathtime," and with commands and requests like "Give it to Daddy" and "Where's the block?" At first, as Church (1966, pp. 56–61) pointed out, the child does not understand what the words themselves mean but rather what the person using them means. The child's own babblings gradually become linked to objects and situations. While nouns generally predominate in the early speech of young children, they are soon capable of using verbs, adjectives, and adverbs to refer to meaningful aspects of their experience (Church, 1966, p. 63). Thus, hot may initially designate any dangerous situation, rather than specifically a very warm object.

The next step comes with the combination of words into utterances, first two at a time and soon more. As psycholinguists have observed (Chomsky, 1968), the composing of original statements means that in learning language, the child has not merely acquired a stock of words, he has also applied the "rules" of grammar and syntax and displayed his ability to construct new and meaningful utterances. The child acquires general principles independent of vocabulary, and this learning does not take place simply by direct imitation. The words have become part of a broad and interrelated structure of meaning and can be actively used in independent thinking and communication.

Church (1966, p. 67) has discussed some of the temporal aspects of early language. The child's first references to past and future are initially couched in the present tense, although the past and future tense forms appear soon after. While it appears that the toddler has vivid recollections of isolated past events, it seems evident that his memories lack continuity, and it is usually not until the school years that the past begins to cohere. The toddler likewise has some sense of the future, but

it probably is even less well articulated than his past. He is likely to
anchor all temporally distant events, past or future, with a single term,
such as *tonight*. Words referring to standard intervals of time (like *hour,
day, year*) or points in time (*last week, yesterday, tomorrow*) may be
employed but without stable meanings. By the age of 2, however, the
child is capable of utterances that suggest that he can plan his own
short-range affairs and order them into workable sequences. And as is
true throughout development, he can understand more complex temporal statements than those he utters.

Very full reviews of observational studies on the development of
temporal concepts as revealed in children's speech have been made by
Paul Fraisse (1963, pp. 151–163, 177–182) and Leonard W. Doob
(1971, pp. 151–163, 177–182). The latter correctly points out that a
serious limitation of this work is that, generally, subject samples have
been very small and almost entirely composed of advantaged children
attending special clinics or schools in Britain, France, Switzerland, and
the United States. Thus, most investigators have overgeneralized in stating that "the child" exhibits a particular feature of temporal awareness
at a certain age. Also, research procedures employed in the various
studies are quite diverse and it is difficult or impossible to compare findings. Recognizing these limitations, we first look at one of the most
informative of such studies and then consider some of the fairly well-
established general findings.

One of the most detailed reports based on systematic observations
of the development of the sense of time in children was provided by
Louise Bates Ames (1946). Using as subjects children who attended a
nursery school at Yale University and were in the high average to very
superior intelligence levels, she sampled records of all spontaneous verbalizations involving or implying time in small play groups that ranged
in age from 18 to 48 months. In addition, at 6-month intervals, the
same children were individually asked a series of questions dealing
chiefly with various aspects of the concept of time. And a second set of
questions about time was asked of older children, ages 5, 6, 7, and 8
years.

In spontaneous verbalization, Ames found that statements indicating present time came first developmentally; then came statements indicating future time and finally statements denoting past time. From 18
through 30 months, by far the greatest percentage of time-related words
used dealt with the present, a few with the future, and almost none with

the past. With increasing age, the relative proportion of present-oriented statements decreased consistently but still predominated; future-oriented statements increased moderately; and past-oriented statements, which did not appear until 24 months, increased slightly. In general, *tomorrow* and other future phrases and tenses were consistently used before *yesterday* and phrases and tenses referring to the past.

Mastery of a particular time concept was not acquired all at once. There were successive levels of attainment: first, the child was able to respond suitably to a temporal expression; next, he was able to use it himself in spontaneous conversation; and eventually, he was able to answer correctly questions relating to the concept. Thus, a child might respond, by waiting, to the phrase "pretty soon" at 18 months; might use the phrase himself spontaneously at 24 months; and at 42 months might answer the question, "When is mommy coming for you?" with the response, "She'll come pretty soon."

For nearly every question asked about time, the children initially gave adequate answers in terms of specific activities and events long before they could give abstract clock-time answers. These included such questions as: "When do you go to bed (have supper, go to school)? When does afternoon begin?" Similarly, rather elaborate expressions of temporal order and sequences might occur as early as 30 months, but correct usage of words implying duration did not generally appear until 36 months.

We present here a few highlights from Ames's (1946, pp. 123–125) detailed presentation of the general findings for specific ages. At 18 months, the children mainly lived in the immediate present and showed little if any sense of past and future. They did not use time words, but they responded to the word *now*. They were impatient and found it difficult to wait.

At 24 months, an important advance was evident. Though the children still lived very much in the present, several words that denote future time had become part of their own spoken vocabulary, particularly *gonna* and *in a minute*. They would wait in response to such words as *wait* and *pretty soon*. They used several different words to indicate present time: *now, today, aw day, dis day*. Though they used no specific temporal words implying past time, they were beginning to use the past tense of verbs, often inaccurately.

At 30 months, the children freely used words implying past, present, and future time. *Morning* and *afternoon* were now used to talk about the present. The future was indicated by *some day, one day, tomorrow,* and several others. There were a few words used regarding the past, with *last night* quite common, though *yesterday* had not yet appeared. The names of days of the week were used freely.

By 36 months, most of the common time words were in the children's vocabularies. There was nearly as much spontaneous verbalization about past and future as about the present, though there was a greater variety of future-oriented than past-oriented expressions. Expressions of duration (*all the time, all day, for two weeks*) came in at this age. There was pretense of telling time and spontaneous use of clock time phrases, usually inaccurate. Children could now tell how old they were, when they went to bed (in terms of some other activity), and what they would do tomorrow, at Christmas, in the winter.

By 42 months, past and future tenses were used accurately. There were many complicated expressions of durations, such as *for a long time* and *a whole week.* There were many new ways of expressing sequence. Children could express habitual action, as *on Fridays.*

At 48 months, Ames reported, many new time words and expressions were added: *when I was little, usually, springtime,* etc. The word *month* was used in many different contexts. Also, such broad time concepts as *next summer* and *last summer* were used accurately. By this age, the children appeared to have a reasonably clear understanding of when the events of the day took place in relation to each other.

Citing Ames's research and a half-dozen similar studies with older children, Fraisse (1963, pp. 179–180) summarized the approximate ages by which the majority could understand and correctly use terms designating location in time. The ages for the accomplishments were: recognize special day of the week like Sunday (4 years old); tell whether it is morning or afternoon (5); correctly identify the day of the week (6), the month (7), the season (7–8), the year (8), and the day of the month (8–9). Fraisse observed that the child first becomes oriented to cyclic activities that bear a direct relationship to the rhythm of his existence, adapting to the regular pattern of every day long before he knows the special days of the week. Subsequently, he masters temporal relationships through the organization into sequences of periods of time he has experienced and orients himself and locates one moment by its relationship to others.

Thus, using the language and concepts provided by his society, the child gradually acquires and incorporates the schemata of time and comes to relate them meaningfully to his own activities and life. Later, we have more to say both about the influence of cultural conceptions of time and the relevance of a temporal framework to the sense of personal identity. But let us again turn to Piaget for a look at other aspects of temporal experience and conceptualization in children.

8. The Development of Conceptions of Time in Piaget's Preoperational and Concrete Operational Stages

Piaget's profound inquiries have shown that children's ideas regarding time are neither innate nor simply learned but rather are intellectual constructions that emerge and undergo successive transformations through experience and action. As anyone who has read Piaget's revealing inquiries realizes—and can readily corroborate for himself—young children have great difficulty in attaining a unified and coherent conception of time. They show many misunderstandings and misconceptions that besides being "cute" and "amusing" to the adult or older child, indicate that our "natural" and "taken-as-evident" notions of time are really highly abstract and elaborated conceptions attained with considerable difficulty.

Piaget's studies of temporal concepts in preschool and school-aged children have particularly focused on two concerns (Elkind, 1974, pp. 38–49; Piaget, 1966, 1971a; 1971b, pp. 59–75). One is the question of how the child attains the abstract conception of uniform clock time in relation to the relative movements and velocities of physical objects. The second question concerns calendar time and chronology, particularly how the child comes to comprehend the relationship of birth order and relative ages with constant, invariant intervals through time. We review here Piaget's findings regarding the gradual attainment of a universal, uniform conception of time in both areas of investigation.

The first topic involves the basic physical equation that velocity or rate of movement multiplied by time equals the distance traveled. Piaget (1966; 1971b, pp. 59–75) has pointed out that traditional views regarding concepts of speed and time generally regard time as the more basic or primitive notion and regard velocity or speed as the more advanced conception of the relationship between distance and time. However, on the basis of his experiments with children, Piaget's contention is that psychological time depends upon velocity, that time is a coordination of velocities, that is, of movements with their speeds. In other words, velocity or speed, while apparently the more complex notion, is actually more primary, and time defined as coordination of movements is the derived and more advanced concept.

Although there are many similarities in conceptualizations of time and space, Piaget (1971b, pp. 60–61) has noted some important differences. Practically, time is irreversible for, except in thought, we cannot go back in time. Movements in space, however, are reversible: we can actually go back again to point A from point B. Another difference he has noted is that space can be considered apart from its contents, as in pure geometry. But time cannot be considered separately from its contents: there is no pure chronometry apart from events. A further psychological difference is that we can perceive a whole geometric figure simultaneously, but a temporal duration, no matter how brief, cannot be apprehended all at once: once an interval is ended, the beginning can no longer be perceived. Thus, any knowledge of time presupposes a reconstruction on the part of the knower. On these grounds, Piaget has argued that knowledge of space is therefore much more direct and simple from the psychological point of view than knowledge of time.

Through ingenious demonstrations and careful inquiry, Piaget discovered that for the young child, time is not clearly separated from notions of distance and velocity. Observing moving objects, the child assesses time in terms of distance traveled and relative speeds (Elkind, 1974, pp. 39–41). Because they lack the abstract adult conception of clock time determined by an independent, uniform, regulated mechanical standard, children's temporal assessments are confused and disorganized. Lacking this conception, they use observed distances and speeds to assess time.

For example, in one study, preschool children were presented with two small mechanical "snails" that were started at the same place and moment but traveled on parallel paths at different speeds for different intervals. When the "slow" snail was allowed to run longer than the "fast" snail but not long enough to catch up with him, the children had no trouble when reporting which had stopped and which was still moving. But when both had stopped, they had difficulty in reporting which had stopped first. Almost without exception, the preschool children said the "slow" snail had stopped first (which was not the case) because it had not traveled as far as the "fast" snail. Evidently, the children were assessing the time spent traveling by means of the distance traveled.

Or in another experiment, children observed dolls simultaneously enter and leave two side-by-side tunnels of different length (Piaget, 1971b, pp. 63–64). While it is clear that the doll in the longer tunnel must have gone faster, the unanimous reply from Piaget's younger subjects was that the two dolls moved at the same speeds. They admitted that the dolls entered and left the tunnels at the same times and that one had a much longer tunnel to go through, yet they asserted that the two went at the same speed because they emerged at the same time. Under the same conditions but with the tunnels removed, the children said that the doll covering the longer distance went faster because they could see it pass the other. With the tunnels again replaced, the majority of 4- and 5-year-old subjects again reported what they had said initially, that the two dolls went at the same speed because they emerged at the same time.

A further experiment (Piaget, 1971b, p. 64) presented two concentric tracks upon which "cyclists" were moved. Young children did recognize that the outside track covered a longer distance than the inside track. Yet when it was arranged so that two "cyclists" went around the tracks side-by-side, arriving simultaneously at the original point of departure, the children said that the "cyclists" traveled at the same speed because they returned to the same spot at the same time. The fact that the outside track was a longer distance was irrelevant to the children's judgments of speed. The feature that was pertinent to their definition of speed was overtaking, and since the cyclists remained side-by-side, no overtaking had occurred. Thus, the young children had a notion of velocity quite different from the classical measure of space related to a measure of time.

Piaget (1971b, pp. 61–62) has distinguished between *temporal order* as the simple succession of events and *temporal duration* as the length of the interval between events. By means of the kind of demonstrations and inquiries just described, he found that a clear conception of the classical notion of velocity as a relationship between the spatial distance traveled and the temporal duration does not appear until quite late in a child's development, at approximately 8 or 9 years of age, when concrete operational thinking is well established. In the preoperational period, before the approximate ages of 7 or 8, there are intuitions of speed that are not based on the ratio. These primitive intuitions are based on ordinal succession and not durations. As we have seen, Piaget found that if young children observed a moving object catch up with and pass

another object, they would report it as going faster. This primitive
intuition of speed based on overtaking appears to be derived from
ordinal spatial and temporal relations, without involving conceptions of
temporal durations.

In the young preoperational child, time is thus not clearly distinguished from notions of velocity and distance. As the child grows older
and establishes operational thought, the abstract concept of time is
attained by combining his elementary notions of speed and of distance to
arrive at the conception of a uniform motion that is independent of the
variable motions and distances that he observes and that are so confusing.
Acculturated older children and adults think of time in relation to the
constant, uniform movements of clocks and regular, established durations
of seconds, minutes, hours, days, and years. Because young children lack
such conceptions, they rely on distance and velocity to assess time. By the
age of 8 or 9, most of the children studied could distinguish between the
time spent traveling and the distance traveled and thus had a sense of the
existence of a general uniform time apart from the particular events they
observed. It should be noted that much research on topics such as tem-
poral-span estimates and delay of gratification behavior shows the
enormous gains children generally make between the ages of 5 and 10 in
mastering and extending temporal conceptions, but this research also
indicates the role of social and cultural influences in such learning (Doob,
1971, pp. 213–272; Rozek, Wessman, & Gorman, 1977).

Piaget was interested to see whether the results from the experiments concerning velocities and clock time held for children's understanding of chronology and calendar time. His studies (Elkind, 1974,
pp. 41–48; Piaget, 1971b, pp. 219–250) of children's ideas about age
and birth order showed similar kinds of initial confusion and gradual
attainment of conceptual clarity regarding time. The preschool child
knows whether he is older or younger than a particular sibling but cannot tell whether he was born before or after the sibling. To be "bigger"
means to be older, but when both siblings are "grown up" it is assumed
that they will then be the same age. Parents and other adults, being
already grown up, are not recognized as aging further, and so on. For
the preschool child, "older" and "younger" do not relate to a uniform
time progression that holds for all individuals. Only when they attain
the abstract reasoning powers consolidated in the operational stages,
beginning around ages 7 or 8, can they deduce from knowing who is
older who was born first.

Thus, workable concepts of calendar time are achieved at about the same age that children attain workable concepts of clock time, both presupposing an understanding of a general uniform velocity determining the standardized measurement of temporal intervals. However, even though the child may have an adequate practical conception of time measurement by ages 7 or 8, he still appears to lack an elaborated sense of personal and historical time. It may not be until well into adolescence, with the great cognitive advances in relational thought and abstract possibilities, that young people can engage in meaningful long-range planning or arrive at a broad historical perspective. And, we may note, even many adults are rather deficient in these advanced attainments. Also, as is discussed later, most other cultures and earlier historical periods have not possessed the heightened temporal awareness and concern with timing that characterizes the modern world.

9. Formal Operations, Ego Identity, and the Extension of Temporal Perspective in Adolescence

Having examined the growth of temporal awareness in early and middle childhood, we next consider the advances occurring in later childhood and adolescence. As noted, by the age of 8 or 9 in Piaget's concrete operational stage, many children show good understanding of and can correctly employ the main temporal terms and concepts in general use. Also, studies in various societies consistently show marked increases in ability to delay gratification and extend temporal span between the ages of 5 and 10 years—though these trends are found to be influenced by intelligence and socioeconomic and cultural factors in varying degree (Doob, 1971, pp. 244–247).

According to Piaget, during the concrete operations stage (approximately 7–11 years of age), children develop elementary forms of reasoning and can solve simple problems without resorting to trial and error. The school-aged child is able to formulate simple hypotheses and explanations concerning concrete matters. He begins to extend his thinking from the actual to the potential and is increasingly capable of dealing with the properties of objects and of relations among them. However, concrete operational thought remains essentially attached to directly observable, empirical reality and does not go far into the envisioning of possibilities. The child in this stage has great difficulty in dealing with hypothetical or contrary-to-fact propositions.

A major feature in the development of formal operational thought, beginning about the age of 12, is that the adolescent becomes increasingly able to consider and imagine possibilities, rather than being confined to the directly observable. The adolescent gradually becomes capable of hypothetico-deductive thinking, in which deduction is no longer confined to immediately perceived situations but extends to hypothetical statements. This major development and the related features of formal operations make adolescent thought much richer, broader, and more flexible than former modes. While the younger child tends to immediately accept as true his initial possible explanation of a problem situation, the adolescent is more capable of recognizing the arbitrary nature of hypotheses and to systematically consider and verify various alternative explanations before finally adopting one.

Formal operational thought not only means that the young person can consider the possibilities in situations and entertain a variety of hypotheses, it also enables him to conceptualize his own thought and to take his own mental constructions as objects and reason about them. Elkind (1967) has insightfully discussed some of the personal consequences of these important intellectual changes upon adolescents' self-conceptions and social relationships. They are a basic feature in the heightened self-awareness found in this important transitional period. These cognitive advances strongly influence emerging personality characteristics and defense mechanisms; the planning of future educational and vocational goals; growing concern with social, political, and personal values; and the developing sense of ego identity (Conger, 1973, pp. 161–166).

With his newfound ability to entertain hypotheses and theoretical positions, the adolescent can both grasp the immediate and imagine various possibilities. Typically, adolescents question previously accepted familial, social, political, and religious beliefs and consider alternative value systems. Many love to conceptualize, reason abstractly, and argue about hypothetical situations. Though such disputation sometimes proves trying to adults, it represents an important exercise of the adolescent's newly acquired ability to transcend imaginatively the limitations of the here-and-now. It seems a necessary stage in his attempt to relate past, present, and future in a personally satisfying and meaningful pattern. Understandably, Aristotle, Rousseau, and a host of later writers on adolescence have in various ways characterized it as the dawn of "the age of reason" and meaningful personal choice for the individual (Muuss, 1975, pp. 9–24).

These cognitive advances greatly affect the changing attitudes toward the self so characteristic of adolescence. The self-concept, of course, has its origins in early childhood and continues to develop and change throughout childhood, adolescence, and adulthood, as many theorists have emphasized (Allport, 1961, pp. 110–138; Gergen, 1971; Hall & Lindzey, 1970; Sullivan, 1953). However, concern with personal identity is particularly heightened in adolescence for many individuals who become more introspective, analytic, and self-conscious.

A central problem for contemporary adolescents is achieving a workable and satisfying solution to the problem of self-definition with its many complicated ramifications. In recent decades, analysis of the adolescent's search for personal identity has become a major theoretical, clinical, and research concern, inspired particularly by the work of Erik Erikson (1963, 1968). In his expansion of psychoanalytic theory, Erikson has emphasized the concept of "ego identity." The adolescent or adult with a strong sense of ego identity feels himself a distinctive individual in his own right, with a satisfying sense of integration. This integration is based on self-perceived and self-accepted consistency over time: a rewarding wholeness through progressive continuity between what the person was, has come to be, and promises to become. Failure to achieve this integration and continuity of self-images Erikson has termed "identity confusion." In Erikson's view, the problem of identity versus identity confusion is the central problem of adolescence. With puberty, the sameness and continuity based on earlier identifications are questioned because of the variety of major changes being experienced. Confronted with the intellectual, social, and vocational challenges soon to come with adulthood, many adolescents are intensely concerned with questions regarding how to connect the roles and skills cultivated earlier with future demands.

The unification of past, present, and future obviously hinges upon a meaningful structure of personal time, and the need for this synthesis is appropriately a central feature of the adolescent identity crisis as characterized by Erikson (1968, p. 181; Gallatin, 1975, pp. 196–198, 214–219). To formulate a coherent plan for his adult life, the adolescent must examine what he is and clarify what he would like to become. In Erikson's terms, an adolescent who is successfully achieving such integration has a sense of "temporal perspective," while one who is failing to do so experiences "time confusion."

A number of studies are in accord with Piaget's and Erikson's views that major changes in the awareness and sense of time take place in adolescence (Cottle & Klineberg, 1974, pp. 70–101; Gallatin, 1975, pp. 214–219). In his major review, Fraisse (1963, pp. 277–280) contended that a full conceptual understanding of time does not usually emerge until relatively late in adolescence—around age 15 or 16. As the adolescent becomes more aware, he generally becomes more concerned with the allocation of time during his own life span. He attempts to imagine his own possible future and actually plan for it.

There are, of course, great differences in the degree to which individuals realistically plan their future and effectively utilize their time (Blatt & Quinlan, 1967; Dickstein & Blatt, 1966; Epley & Ricks, 1963; Kastenbaum, 1961; Le Shan, 1952; Wessman, 1973). Despite a good deal of research on the socioeconomic and personality correlates and determinants of these differences in temporal organization and experience, the overall findings unfortunately are equivocal and often contradictory (Doob, 1971, pp. 255–272). Some of the possible reasons for such confusing findings are considered in a subsequent chapter.

However, the general notion of a marked expansion of temporal awareness and an increase in personal planning during adolescence is clearly supported by a number of investigations. Douvan and Adelson (1966) found 14- to 18-year-old American adolescents developing increasingly specific sets of expectations and aspirations. The boys generally had rather well-organized and realistic views of the future, and the older they were, the more evident this orientation became. The girls' plans were generally less well articulated and hazier, probably reflecting cultural sex-role socialization. Monks's (1968) study of 14- to 21-year-old Dutch adolescents found similar results. Writing essays on their future plans and aspirations, late adolescents generally produced much more complex and differentiated sets of temporal perspectives than did early adolescents. The major writers on vocational development (Borow, 1976; Ginzberg et al., 1951; Ginzberg, 1972; Havighurst, 1964; Super, 1957) all emphasize the increasing realism and structure in temporal orientation and life plans that commonly occur in adolescence and the early twenties. In a later chapter, Thomas J. Cottle will present some vivid examples of how pressing these temporal concerns are for some young people.

10. Temporal Concerns and Perspectives in Adulthood and Aging

Despite the fact that the adult years constitute the greater proportion of the life span for most individuals, until fairly recently the period was greatly neglected in psychological theory and research. Most developmental accounts were almost exclusively concerned with the formative periods of childhood and adolescence and, with a few notable exceptions, simply ignored the long and very full years of adulthood. Happily, this imbalance has begun to be corrected. Though much work still needs to be done, there has been increasing recognition and investigation of the many critical developments and changes occurring throughout the mature years. Many important transitions are inevitably time-related because of their intimate connections with the psychosocial dynamics of growing old. A much richer understanding of changing temporal perspectives, orientations, and concerns should eventually be a product of this increased attention to adulthood and aging.

Because of the individuality and the variability in personal life histories and because of the practical difficulties in the execution of well-designed longitudinal studies, broadly based and well-documented general findings have been slow to emerge. We indicate here some major contributions to this expanding field that have clear relevance to an understanding of changing temporal perspectives in adulthood.

Charlotte Bühler (1933, 1968) was a pioneer in collecting and analyzing biographical life-history data on creative and ordinary individuals. Based on detailed study of over 200 cases, she discerned five general phases and their approximate ages that characterize many normal life histories. The phases were described as: (1) before the self-determination of life goals, from birth to age 15; (2) tentative self-determination of life goals, from age 15 to about 25; (3) more specified and definitive determination and implementation of goals, from age 25 to about 45 or 50; (4) assessment of the foregoing life and its relative degree of fulfillment from ages 45 or 50 to age 60 or 65; and (5) a final phase of rest and retirement from ages 65 or 70 to death. Bühler emphasized the "intentionality" of directed strivings toward the attainment of personal developmental goals that gradually emerge and find relative degrees of fulfillment or failure in the individual's life. These goals are realized in various activities, personal relationships, and feelings about the self. Bühler's accounts are rich and detailed and have great interest for their consideration of temporal orientations and perspectives at the different stages.

Bühler's (1968, pp. 73–74) view is that it is in the second phase, between ages 15 and 25, that there is generally "a first grasp of the idea that one's own life belongs to oneself and represents a time unit with a beginning and an end." During this period, the life goals are conceived of tentatively and experimentally, and the young person begins to see himself in historical perspective. In the following adult period, from 25 to 45 or 50, the life goals are set and implemented in more specified and definitive ways. It is Bühler's view (1968, p. 74) that the healthy, self-realizing person lives predominantly in the present, seeing his future tied to his present life in meaningful continuity. The immediate future is closely intertwined with the present and necessitates realistic planning and action. The more removed intermediate and distant future may be planned with varying degrees of flexibility or specification. Bühler, of course, has recognized and discussed the various impediments, obstacles, and setbacks that may make it difficult for many individuals to achieve such harmonious and integrated temporal perspectives.

Bühler found the climacteric years from about 45 to 65 a period in which most people have to reorient themselves for many reasons. Self-assessment includes a stocktaking of the past and revised planning for the future in light of necessary limitations. There may also be anticipatory thinking about the last years of life. Relative feelings of fulfillment and unfulfillment are strong concerns in this period.

Later life after 65 shows great individual variability, with some people undergoing great decline and marked restriction of activity, while others remain active and involved. Health and general life circumstances are obviously of great consequence. Bühler's impression was that relatively few people enjoyed feelings of essential fulfillment in the last phases of their lives. While the incidence of actual despair seemed small, the pattern of the majority seemed best described as "resignation."

One of the major studies on middle life and aging was carried out in Kansas City during the 1950s by Bernice L. Neugarten and associates (1964). During a 7-year period, they studied over 700 individuals in good health using interviews and projective tests. The study included both cross-sectional and longitudinal data on personality change in adults. A major age-related change was a shift to "increased interiority," beginning in middle age and becoming more marked in later life, with the individual becoming increasingly self-reflective and introspective and less concerned with the external social environment.

In a related study of 100 high-status men and women, Neugarten (1968, pp. 93–98) examined changing time perspectives in middle age. Both sexes, particularly the men, talked of a difference in the way time was perceived. Generally, the middle-aged viewed themselves as a bridge between generations, both within the family and in the wider contexts of work and community. At the same time, they had a very clear sense of differentiation from both younger and older generations. Their memories of particular historical events and their general accumulation of experience created a generational identification and separated them from those who were younger. They felt increased proximity to and identification with those who were older, and they were likely to reassess their relationship with and feelings toward their parents. Both sexes talked of a difference in the way time is perceived, in that one's life becomes "restructured in terms of time-left-to-live rather than time-since-birth" (Neugarten, 1968, p. 97). Related to this shift in temporal orientation in middle age was an increased awareness that personal time is limited. The death of contemporaries is a particularly sobering reminder that there is "only so much time left," which was a conspicuous theme in the interviews. The recognition that death is real had much more immediacy than it had formerly had. Yet most of these advantaged middle-aged respondents felt that middle age was "the prime of life," a period of maximum capacity and ability to handle a highly complex environment and a highly differentiated self. Very few expressed a wish to be young again. Instead, most appeared content with their present capacities and felt that their competence, judgment, and better grasp of realities were reassuring aspects of being middle-aged. Stocktaking, heightened introspection, the structuring and restructuring of experience, and a sense of one's priorities and competencies seemed to characterize this sample of individuals, who were obviously doing well in mid-life.

One of the controversies regarding the psychological characteristics of older age stems from Elaine Cumming and William E. Henry's (1961) formulation of "disengagement" as a general feature of aging. This theory postulates that the aging individual gradually withdraws from his society. In the later years, the individual recognizes that his future is limited and that there is insufficient time to do what he hoped and planned. Anticipating death, the individual constricts his life plans and activities. Evaluations of disengagement theory (Kastenbaum, 1969; Kimmel, 1974, pp. 314–317; Neugarten, 1968, pp. 159–172) indicate that while such patterns of increasing constriction clearly characterize the later years of many individuals, there appear to be many individuals whose involvement in life continues to be high. In a later chapter, Robert Kastenbaum discusses some of his observations on the very varied temporal perspectives found in old people.

While much remains to be done, certainly the increased research and theory concerning adulthood and aging is promising. Both popular and scholarly accounts are giving greater attention to the adult years and are attempting to formulate the major developments and phases (Davitz & Davitz, 1976; Erikson, 1963, 1964; Gould, 1972; Kimmel, 1974; Levinson *et al.,* 1974; Maas & Kuypers, 1974; Neugarten, 1968; Roff & Ricks, 1974; Sheehy, 1976). As is amply evident by now, throughout life one's personal experience and structuring of time are undergoing development and change.

Having considered many of the outstanding features of changing temporal awareness and orientation throughout the life cycle, we now wish to shift attention to some other important features of the personal experience of time. We wish to examine its relationship to memory, to the phenomenology of consciousness, and to cultural and historical awareness. Our main concerns are some of the ways that normal adult experience is structured in time.

11. The Active, Integrative, and Personal Character of Memory

A psychological examination of time inevitably leads to a consideration of memory. Temporal awareness and memory are intimately related in human experience. Aristotle, the first significant figure in psychology and the father of Associationism, recognized this close connection in his essay "On Memory and Reminiscence," in which he observed that "all memory implies a time elapsed, consequently only those animals which perceive time remember." Personal recollection includes consciousness of "formerly" and the distinction of "former" and "latter" (Aristotle, quoted in Straus, 1966, p. 60).

Throughout the modern history of psychology, memory has been a constant topic of theoretical concern and research activity. It is far from fully understood, with many issues still unresolved and subject to continuing inquiry. We cannot review all the relevant work here. However, it is clear that recent decades have seen many theorists conceptualizing remembering as far more of an active constructive and reconstructive process, instead of the fairly passive registration and reactivation of "ideas," "traces," or "engrams" of stimuli and events postulated in most earlier accounts.

The dynamic features of memory were recognized by Sir Frederic Bartlett, who emphasized the role of cognitive schemata in organizing memory. Bartlett's (1932) experimental findings led him to conclude that remembering is an active process involving the reception, categorization, and reconstruction of material according to idiosyncratic organized patterns. He argued that in remembering, as in other cognitive tasks, the individual first attempts to find sense and meaning in material presented to him by relating it to what he already knows about the subject or ones similar to it. Such preexisting knowledge, which selectively categorizes and alters past and present experience, was termed *mental schemata* by Bartlett. He held that schemata were the basis not only of memory but of concept formation, problem solving, attitude formation, and a variety of other cognitive activities.

Bartlett held that individuals refer all incoming stimuli to selective and organized, yet modifiable, mental structures. Recall is a joint function of both input characteristics and the individual's preexisting schemata. Human thinking requires the active reorganization of memory and its combination with new information. Cognitive acts of recognition, understanding, and judgment require the use of memories and complex mental operations organized into persisting, yet modifiable, frameworks. In Bartlett's (1932) words, memory is an "achievement in the line of the ceaseless struggle to master and enjoy a world full of variety and rapid change" (p. 314). Memory, and all the images and words that go with it, is an aspect of the development of constructive imagination and thought whereby man transcends the narrowness of presented time and space.

Current cognitive views of memory emphasize active information-processing involving the construction of complex and integrated organizational schemata (Hunter, 1964; Lindsay & Norman, 1972, pp. 286–434; Neisser, 1967, pp. 279–305; Pollio, 1974, pp. 246–270; Posner, 1973, pp. 1–60). In accord with cognitive information-processing analyses of perception, problem-solving, and concept attainment, these views of memory assume an active subject who is constantly interpreting and organizing his world. Once information has been coded and assimilated in the initial processes of registration and storage, it later has to be reconstructed in order to be usefully retrieved. Storage and retrieval are conceived of as hierarchically organized processes using abstract rules of categorization and complex evolving schemata that both assimilate and accommodate to new experience. These contemporary, dynamic, and integrative accounts of memory processes accord with the constructive and transactional character of coming to know the world that was described in our earlier consideration of the development of temporal awareness.

While knowledge of the general features and the common basic processes of memory organization is essential, we wish to emphasize that the particular organization of memory and the specific memories recalled by each individual are unique and personal. In early childhood, schemata and concepts are only vaguely defined and are not well integrated with other stored information. Understanding is gradually elaborated as personal experience accumulates, as examples and characteristics are learned, and as class relationships and schemata evolve. Later in life, when a great deal of information has been accumulated and organized into a richly interconnected data base, learning takes on a new character. New things can be learned and understood by analogy with and reference to what the individual already knows and remembers. New experiences are assimilated into preexisting memory. The whole of past experience is automatically brought to bear on the understanding of the new events.

Dynamic and integrative views of memory recognize the development of idiosyncratic cognitive organization as the rule rather than the exception (Lindsay & Norman, 1972, p. 432). It is unlikely that any two people would evolve exactly the same conceptual structure to represent the world they experience. While the basic features of memory and the underlying processes for manipulating and reorganizing information are probably quite similar from individual to individual, personal cognitive-construct categories and organized memory systems may be very different. Thus, there undoubtedly is considerable individuality in memory organization. However, as we earlier indicated, language, symbolic systems, and general patterns of thought do tend to organize experience and memory in shared and socially congruent ways. Thus, memory possesses both collective and individual features—as is richly attested to by the uniformities yet diversities in human experience.

The succession of happenings and events in each individual's life is the primary phenomenal basis for remembering and recollecting. The locus of experience is personal awareness, the private and vital "stream of consciousness." Awareness of "now" and recollection of "then" are rooted in our immediate experience and understanding. The writers who have most cogently treated the basic experiential nature of awareness, memory, and time are the phenomenologically oriented theorists. We next consider some of their important ideas.

12. Phenomenological Views of Temporality and the Stream of Consciousness

The major phenomenological theorists, including William James and Edmund Husserl, have emphasized the significance of temporality and continuity of consciousness as basic constituents of adult human experience. Expanding on this view, Aron Gurwitsch (1966, pp. 301–331) has argued that continuity of consciousness is identical with phenomenal time. He noted that consciousness appears unbroken in immediate experience: the succession of events and phases is connected, and even in abrupt transitions there is generally a lingering awareness of what went before. Also, the present moment usually contains some anticipation, however vague and uncertain, of what is to come. In most situations, we have fairly reliable general expectations of what will follow. Besides such transitions and sequences, events with duration seem to possess an intrinsic temporality. When a note resounds, the listener is aware of its duration, recalling its onset and anticipating its continuation or termination. Thus, the immediate "present" of conscious life is pervaded by reminiscences and expectancies.

The temporal continuity and flow of experience was vividly

described by William James (1948, pp. 151–175, 280–301) in his famous chapters on the stream of consciousness, time, and memory. He observed,

> The practically cognized present is no knife-edge, but a saddle-back, with a certain breadth of its own on which we sit perched, and from which we look two directions into time. The unit of composition of our perception of time is a *duration,* with a bow and a stern, as it were—a rearward and a forward-looking end (James, 1890, Vol. 1, p. 606).

While the content within the temporal frame undergoes incessant change and variation, what James called the "time coefficients" themselves—the "actual now," the "not yet," the "just gone"—seem to persist as enduring references in immediate experience. Thus, no mental state, however elementary, is entirely confined to the "present"; none is without some features pointing to the past and the future.

In normal adult experience, it appears that consciousness is necessarily an awareness of time passing. In sequences of awareness and action, consciousness is carried along unbroken. Thus, continuity inheres in consciousness, and it is experienced as a flow or stream. Continuity and temporality are then two aspects of the fundamental structure of conscious life.

Such phenomenological analyses of experience find temporality to be an intrinsic aspect of normal adult consciousness. At every moment, consciousness appears in continuous flux. Mental life possesses form and meaning by virtue of its temporal coordinates and organization. For consciousness of an atemporal nature, there could be no identity, no identical object, no recollection or anticipation. Time appears a necessary condition of consciousness.

These phenomenological ideas are further developed in the work of Erwin Straus (1966), particularly in his insightful discussions of sensory encounter and motor activity. Straus discussed how, when awake and conscious, one feels in immediate contact with reality. Upon awakening, we discover that we have slept and distinguish between being awake now and having dreamed last night. The well-ordered temporality of existence resumes. Unlike the disconnected and bizarre time of the dream, waking hours move fairly regularly and predictably, with measured pace and continuity from one event and activity to another. Waking events fit into the continuous narrative of our life histories, as dreams do not. While dreams are not truly timeless, the order of time dissolves, and they seem incoherent and apart from our structured daily existence.

Upon awakening, we experience the beginning of a new day. Waking finds its place in the order of world time, as Straus (1966, pp. 105–106) cogently described. The moment we arouse from sleep and "come back" to ourselves and to the world, we usually wonder, "What time is it?" With this question, we attempt to relate the actual present, a personally experienced "now," to the embracing temporal order of the world. We locate our "here-and-now" in a broader conceptual structure of space and time, extending awareness beyond the borders of the present moment.

Upon awakening and throughout the day, we locate ourselves in a specific here-and-now within the established abstract frames of space and time. Self-awareness and awareness of the world are intimately linked together in normal adult experience. As we are active, living creatures, reality is not detached from us. It is simultaneously imposed on and constructed by us through our perceptual, sensorimotor, and cognitive operations. Goal-oriented actions are directed to objects in the environment. Our bodies, senses, and actions place us in immediate relation to the world. The present location and time, always a personal here-and-now, are a transition point en route to other places. Insightful, directed activity requires conceiving of an extended spatiotemporal field, proceeding from given positions toward intended destinations. Personal action altering the *I–world* relationship is actualized by means of a temporal horizon open to the future. The present moment becomes subordinate to the future and the past in the span of personal time. Activity can follow a plan. By freeing himself from the constraints of the immediate situation, man gives meaning to his experience and becomes responsible to himself (Straus, 1966, pp. 166–187, 210–216).

The phenomenal experience of time is not merely a series of discrete and disconnected "nows"; rather it generally unfolds in continuing, interrelated, and meaningful sequences. Speech to the listener is not just a series of acoustic vibrations, phonemes, or even words; it is apprehended as communication. The words are understood as part of a whole deployed in time. The first word opens the sentence, pointing forward to those yet to come. The grammatical subject will be defined by the predicate that follows, the end referring back to the beginning. Both speaker and listener apprehend this temporal sequence of words as a whole with definite meaning.

Events are seldom experienced as fragmentary and disjointed; we customarily apprehend interrelated and meaningful wholes through the mediation of imagery, language, and symbolic representation and transformation. Spatial and temporal schema relate our immediate here-and-now to the locations of yesterday, today, and tomorrow in our own lives and the shared social frameworks for understanding.

We turn again to consider memory, but this time with a more phenomenological emphasis. Straus (1966, pp. 59–74) contributed many insightful observations about the personal experiential character of remembering. He noted that remembered objects, events, or persons are not actually present in the immediate "now" but rather are present *in absentia*. They are "re-presented" in the act of remembering. The past gains its meaning through its relation to and distinction from the present. While the objective relationship "earlier" or "later" between two events remains invariant, the point of division between "past" and "future" is the subjectively experienced personal "present." Past and present are temporal qualities related to the personal order of time. Remembering occurs within the frame of historical time and of one's own life history. Recalling the past, one is aware of the continuity of personal existence and refers to the historical self. In the personal order of time, the past is maintained with meaning and significance.

Straus (1966, pp. 66–67) argued that memories are not passive registrations or exact copies of the original events as some theories of memory "traces" assumed. The transition from original perceiving to remembering is not a change characterized just by loss of details but is an essential transformation. Completely faithful and full memory would be a dubious blessing, for it would reproduce events in every detail, like speech played back on a tape recorder. Memory selects and condenses. Events are not customarily remembered in their original fullness and character. For example, knowing the outcome of a former uncertain and tense situation necessarily alters its recollection and changes its emotional character. We may still remember that we were excited but can hardly reproduce the original excitement. In recalling past joys or sorrows, we usually contrast them with the present and thereby gain new perspectives on the past. We may remember the past, but we cannot bring it back.

Straus (1966, pp. 67–74, 290–295) discussed the autobiographical features of remembering as an account of one's life history. He distinguished two aspects of personal time: the narrow repetitive cycle of immediate mundane needs and the extended perspective of the historical self. We generally remember few details of the banal daily routine once it is past and done. When forgotten, such oblivion does not seem to be the result of repression. Straus has held it unlikely, in contrast to some psychoanalytic writers (Rapaport, 1950, p. 141; Whitrow, 1963, pp. 89–110; 1975, pp. 26–38), that every event is permanently stored in memory and potentially recoverable. In the routine activities of life, the happenings of one day seldom have particular meaning or importance. Little remains to be remembered because nothing important or distinctive was registered in its particularity.

Straus's view is that significant and novel happenings are the ones most likely to be disengaged, arrested, and therefore registered and later remembered in personal recollection. The memorable event is different and distinguishable from what has been experienced or known before. The personally memorable belongs in a historical context, occurring for the first time in one's personal history or in that of one's group. What is notable is determined not only by its contrast with the past but also by its significance for the future. Emerging from the current of neutral events, important occasions are incorporated into the whole of personal memory according to their content, their temporal position, and their historical weight. Every experience receives its specific significance and value from its temporal position in the individual's life history.

The individual also relates his personal "now" to the objective, common, general order of time determined by calendar and clock. However much the constructors of various calendars and clocks may have differed in how they divided days, months, or years, or in the instruments they used to measure time, they agreed on the basic idea of a cosmic order of time that is one and the same for everybody and, therefore, indifferent to the fate of the single individual. Clock time flows uniformly, with one day as long as another, and it is quantifiable, being strictly determined by measure and number. Personal time, however, is not homogeneous; in it, one year is not necessarily as long as another one. Neither is it determined simply by measure and number, but rather it is articulated by values and the importance or unimportance of events. There is a profound contrast between the cosmic objective order of measured time and the individual time of our personal experience.

The objective–subjective contrast is experienced as accordance or discordance of these two temporal orders. In normal experience, time may appear to pass rapidly or slowly. On a boring day, subjective time passes slowly, but in retrospect the day seems short. On an eventful day, subjective time passes rapidly, but the interval between morning and night appears long. And in many forms of psychopathology, temporal discordance, disorientation, and fragmentation are manifest—as we discuss more fully in a later chapter. The present is not experienced as meaningfully connected with past and future. The personal relation to the objective order of time becomes distorted. With severe arrest and breakdown in a patient's life, experienced time may be transformed into a mere sequence of events that, finally becoming meaningless, loses even the character of temporality.

But in normal everyday experience, personal time maintains its general coherence and order. Past, present, and future relate meaningfully in the contexts both of one's personal life and of the temporal framework established by one's society. Having examined the phenomenological structure of personal time, let us now turn to the cultural and historical aspects of temporal awareness.

13. Cultural Conceptions of Time

It is impossible, of course, to state with certainty when and how full conscious awareness and elaborated conceptions of time first arose in human society. The evolutionary development of the brain, particularly an increase in the frontal association areas of the cerebral cortex, appears to be a necessary substrate for man's advanced temporal awareness. Marked advances in cranial capacity occurred sometime during the past million years, possibly earlier, according to the fossil evidence (Maxwell, 1971, pp. 37–43). Many of the significant discoveries and practices of early man clearly required foresight and planning or indicate considerable temporal awareness and concern, for example, tool making, fire making and tending, agriculture and settled habitation, and burial customs. These prehistoric practices must have been accompanied by the development of social communication and speech, which permitted the maintenance and transmission of cultural practices and traditions. Language facilitates memory and enhances capacity for imagination and planning, thereby extending the time span into past and future.

Apparently, all four thousand or so known languages enable their speakers to designate temporal relationships and to distinguish between

past, present, and future events—though with varying degrees of difficulty (Maxwell, 1971, p. 42; Stutterheim, 1966, pp. 164, 170–171).
Presumably, the speech of our prehistoric ancestors provided for similar
distinctions. Provocative interpretations by certain anthropologists and
linguists argue that conceptions of the temporal features of "reality" are
structured very differently or may even be incapable of being expressed
in the linguistic systems of various cultures (La Barre, 1954, pp. 187–
207, 361–362; Lee, 1959, pp. 105–120; Whorf, 1956). However, it
seems to be the consensus of most informed scholars who have carefully
reviewed this matter that the more extreme assertions of linguistic
relativity, though challenging, are far from established (Black, 1968, pp.
90–95; Carroll, 1964, pp. 106–111; Church, 1966, pp. 132–136; Maxwell, 1971, pp. 43–47). Certainly most, and possibly all, languages
possess time words and allow their speakers to communicate regarding
temporal features of experience. Also, context and paralinguistic features probably would allow implicit temporal references that might not
be clearly codified in speech. We doubt that any group could function or
survive without some degree of effective communication regarding the
temporal features of both the natural world and social interaction.

Yet clearly, there are many very important cultural and historical
differences in temporal awareness, conception, and orientation. These
differences are often subtle and difficult to discern because they are
implicit assumptions of the entire world view and consciousness of a
particular culture or period, pervading many aspects of thought and
action. Further, because they touch on so many features, they are hard
to classify and organize systematically. Some of the significant aspects of
cultural differences in temporal features and orientations involve
dominant images and concepts of time; elaborated methods and techniques for time keeping and reckoning; an awareness of tradition and
history; static versus dynamic conceptions of nature and society; a relative orientation and attitudes toward past, present, and future; a degree
of scheduling, planning, and social organization of activities; and a
general awareness and concern regarding time and timing. The sources
and consequences of cultural and historical differences in temporal
experience and awareness are a rich and complex area of inquiry that,
in our opinion, has yet to receive the attention it merits. Some stimulat-
ing accounts have treated various sociocultural and historical aspects in
depth or have suggested the range of possible inquiries (Hall, 1973;
Kluckhohn, 1953; Kluckhohn & Strodtbeck, 1961; Maxwell, 1971;

Nakamura, 1966; Needham, 1966; Russell, 1966; Sorokin & Merton, 1937; Whitrow, 1975; Zern, 1970).

Full realization of the unusual degree of temporal concern pervading and regulating almost every aspect of modern life and consciousness is both illuminating and disquieting (de Grazia, 1964, pp. 289–312; le Lionnais, 1959). Certainly, it appears to be without precedent. Though some sophisticated calendars and rudimentary methods for keeping time existed in earlier periods and cultures, accurate time-keeping and clocks began to become widespread in western Europe during the late medieval period (Fraser, 1975, pp. 47–67). The mechanical clock did not appear until the 13th century and gradually became an important feature of church towers and public buildings. During the following centuries, clocks and watches were successively improved, and their possession became a mark of status for those rich enough to afford them. The great mechanical improvements in the clock and its obvious utility in navigation, science, and commerce made it an object of increasing importance. The fascination and the heuristic significance of the clock for the intellectuals and scientists of the 17th century have often been noted. It became the symbol of a great cosmic order to be found in the natural world, to be rationally understood and used by man—but it eventually came to regulate him!

During the 17th to 19th centuries, a new awareness and concern with time emerged, permeated the expanding industrial world, and became an essential component of the ever-increasing pace of organized and coordinated activities in manufacturing and commerce. But not until the 19th century did clocks truly become a widespread possession of the public at large—mechanisms whereby average people were regulated both in work and in the remainder of their daily lives. Inexpensive watches began to be manufactured in Switzerland in the 1860s, and within a few decades, millions of watches were being produced in the world (de Grazia, 1964, pp. 290–291). In factory and office, activities became increasingly scheduled. Time became a valuable commodity to be saved, spent, and sold. Wages and hourly rates gave the impression that time, rather than labor or services, was what was bought and sold. Clock time first scheduled work time, while outside social and community life initially held to the more leisurely and unregulated pattern. But as de Grazia (1964, pp. 291–302) cogently argued, "free-time" and "leisure" were increasingly dominated and subtly changed by a pervasive awareness of and regulation by the clock.

To become a negotiable commodity, time had to be abstracted and neutralized. Days, hours, and minutes must be regarded as interchangeable like uniform, standardized parts. The former unconcern regarding punctuality and schedule still found in "backward" regions of the world would not serve to coordinate modern industrialized and bureaucratic society. With uniform precision, the clock breaks the day into standardized units that should be filled with worthwhile activities. Time should not be wasted—one has only 24 hours a day. Modern men and women are extraordinarily time-conscious. We lack the freedom from temporal constraint and the easier regulation by natural pace and rhythm that seems to characterize simpler ways of life. In home and school, the modern child observes and is taught the need for punctuality and schedule that will govern the rest of his days.

Repeatedly in our daily routine, we are concerned about the passage of time and consult the clock. Farmers and artisans in earlier periods undoubtedly worked hard, yet it seems that they had less cause to worry about time. But today's factory workers, office employees, managers, and professionals must of necessity be very aware of the time governing their activities. The average worker earns his livelihood by applying himself to tasks set by others and coordinated to their demands and schedules. Routines and plans must be imposed if complex coordinated operations are to be carried out effectively. If one is to be successful in most contemporary occupations, one must be able to allocate and use time appropriately. People and organizations are required to be dependable, productive, and efficient—or suffer the consequences.

The practical mechanized conception of clock time has come to dominate our age. In many contexts, particularly in the realm of work, we regard time as standardized and linear. It does not repeat itself, it ticks off in a straight line from t_1 to t_2 at a steady rate. It is the time that Isaac Newton conceived of: real and mathematical, flowing uniformly, embracing all phenomena but aloof from them, universal and homogeneous with each instant the same everywhere (de Grazia, 1964, p. 303). While advanced science and personal experience may admit relativity; the practical world does not. We are regulated by the Newtonian world of timepieces.

Ever aware of the clock inexorably ticking away, modern man resents its coercion. Pressured and harassed, he longs for the temporal unconcern apparently found in "simpler" ways of life. Much of the counterculture search for Nirvana and rejection of the "rat race" seems to be a rebellion against temporal constraints and pressures. Even the successful organization man, when martinis fail, seeks occasional escape to idyllic Caribbean or Aegean havens where clocks are temporarily forgotten. Probably such pastoral idylls are just romantic fantasies that ignore or distort the harsh struggle for survival that was the lot of most of mankind in "simpler" times. We suspect that is largely the case. For many people, life always seems to have been better "then" than it is "now." But at any rate, it is clear that modern man is extraordinarily time-conscious and that this heightened awareness is often burdensome despite the obvious material advantages it brings.

14. The Modern Awareness of History

A further impressive temporal achievement, though again with its oppressive features, is the modern awareness of history, both cosmic and human. Recognition and appreciation of the vast scope of past history have their own long story that we merely sketch here. Between 800 and 200 B.C. a series of major intellectual developments occurred in the great civilizations of China, India, and the West (Jaspers, 1960). The great philosophical traditions took shape as, seemingly for the first time, men questioned and contemplated their existence and the world around them. They became aware of consciousness itself, and the nature of thought and its limitations became objects of speculation. Basic ideas were advanced and debated that have remained as the classic foundations of the enduring major intellectual, philosophical, and religious orientations. Reason, knowledge, and practical experience slowly began to challenge ignorance and myth. Political change and turmoil shattered the static conditions and conceptions of earlier periods. A consciousness of history and change began slowly to emerge that, though often dormant, has nonetheless persisted to vitalize mankind's creative accomplishments.

Most early cultures appear to have held predominantly cyclical views of time. Western conceptions of linear time gradually took form through the eschatological visions of monotheistic Judaism and Christianity that saw human history as a cosmic drama of sin and salvation (Eliade, 1959, pp. 95–137). Christian thought transformed the mythic themes of eternal repetition and invested both history and personal life with a sense of immediate significance and ultimate destiny. Nonethe-

less, throughout most of the medieval period, both cyclical and linear conceptions of time seem to have coexisted or contended, and the views of time were still predominantly religious, not secular (Whitrow, 1975, pp. 11–25).

From the 17th century on, linear and naturalistic conceptions of time and historical development became increasingly common. The Enlightenment's optimistic faith in infinite progress became popularized in the 19th century, though that belief has undoubtedly been shattered in our own time (Dardel, 1960; Eliade, 1959, pp. 141–162). In any case, the modern period has seen an extraordinary extension of temporal awareness in history and science. A true concept of history did not reach full maturity until well into the 18th and 19th centuries through the works of Vico, Herder, Ranke, and other scholars (Cassirer, 1944, pp. 217–260; Toulmin & Goodfield, 1965, pp. 103–140). Conceptions of historical development based on objective and critical research were slow to develop, but once achieved, they inevitably altered and enriched the conception of the past. The natural sciences during the last four centuries likewise saw the gradual emergence and interrelation of a series of fundamental discoveries and reconstructions of the past in astronomy, geology, paleontology, and biology (Toulmin & Goodfield, 1965, pp. 74–102, 141–231). As with history, it was not until well into the 19th century that these extended time perspectives became securely established and interrelated in the sciences. The evolutionary and developmental synthesis of Darwin was a major culmination of this long sequence of painstaking empirical observation and theoretical reconstruction. It can thus be seen that it is only relatively recently that historical and scientific inquiry has fully recognized the vast extent of past time.

These extended temporal perspectives have transformed modern thought. In the last two and a half centuries, during which the recognized universal time span has expanded from 6,000 to 6 billion years, scholars have been obliged to reformulate all their beliefs in order to fit them into this vast new perspective. Those who are cognizant now have a conception of the universe, life on this planet, and human history that is truly staggering. What scholars and scientists know or attempt to conceive so far transcends our immediate experience that we must necessarily be awed. Our limited personal temporal span seems inconsequential in the vast cosmic perspective. A considerable part of the ideological crises of our day stems from the challenges to and the gradual erosion of long-cherished doctrines and convictions as we attempt to cope with this unprecedented transformation of temporal perspective. In addition, the tremendously accelerated pace of social and technological change gives modern men and women ample cause to feel that they have lost their bearings (Toffler, 1970). In a later chapter, Arthur M. Schlesinger, Jr. will show how profoundly this distressing sense of hurtling through vast uncharted reaches of space and time pervades much modern thought, literature, and art. Our times indeed are "out of joint."

Science and critical thought began in the classical Ionian world with a search for permanent features behind the flux of experienced phenomena. From the beginning, a few philosophers, like Heraclitus, argued that there are no fixed elements to be found apart from the creation of our own minds. All in the natural world is in constant change, while presumed permanence and certainty exist only in the world of the intellect (Toulmin & Goodfield, 1965, pp. 271–272). This doctrine was hard to accept, and through the ages, innumerable thinkers have tried to convince themselves and others that they have established the true and permanent order of things. But the skeptics may have been right. Instead of vainly searching for permanent and defined order, perhaps we are learning to accept and live in an evolving universe and social world where all things are recognized as subject to temporal development and change. The fear that the absence of a fixed and objective order of existence necessarily means intellectual chaos and nihilism may be groundless. We must find and create our own shared systems of human meanings, values, and purposes—which is, in fact, what societies have somehow always managed to do.

Though our heightened awareness of universal history may be disquieting and may make us acutely aware of our limitations, it does seem that at least, we understand the world and its transitory processes far better. Finally, we must be impressed with an intelligence sufficiently self-aware to recognize its own finite limitations and capable of both creatively constructing and finding its own location and significance within the vast frames of time and history.

15. Conclusion: The Character of Human Time

Time has proved extraordinarily elusive and hard to grasp. Through the centuries, it has tantalized and frustrated thinkers attempting to capture its essence and to formulate it conceptually. It seems such a fundamental feature of the world and so integral to human experience; yet time is not tangible and is certainly far from simple. Our inquiry finds that it does not appear to be an elementary feature immediately given in experience and directly apprehended. Rather, the notion of time is a highly abstract mental construct that develops gradually from repeated active encounters with and personal experience in the world and that involves elaborated cognitive schemata based on inner imagery, symbolic representations, and language. The synthesis of these complex relational schemata into our sophisticated temporal awareness and abstract conceptions of time is a high-level intellectual achievement. It is based on the advanced symbolic potentials of the human mind that are realized and elaborated in our culturally and historically shaped systems of thought.

Temporal awareness and intellectual conceptions of time are reflections of what we apprehend as the presumed underlying order and system in the events of the natural world. But the temporalities that we conceive of are inevitably determined by our creature capacities and limitations. Very likely, lower, and possibly higher, orders of life "experience" and "know" time of a very different character, reflecting their particular potentialities for engagement with their environments. And the symbol systems and world views of various cultures and periods have been shown to have very different temporal images, attitudes, orientations, and conceptual frameworks.

The time that we humans know is not just a cold, neutral, objective sequence of happenings in the natural world. It is not something apart and detached, though we can intellectually conceive of it as such. Rather, the time that we experience and that most concerns us is *our* time, personal and social. It is related to the structure and the scope of our lives: the store of past memories, the worries and concerns, the future plans and aspirations. It is a vital personal time, fraught with continuity and significance, that we must grasp and shape in order to create the meaningful patterns of our lives.

References

Abraham, K. *Selected papers of Karl Abraham.* New York: Basic Books, 1927.
Allport, G. W. *Pattern and growth in personality.* New York: Holt, Rinehart and Winston, 1961.

Ames, L. B. The development of the sense of time in the young child. *Journal of Genetic Psychology*, 1946, *68*, 97–125.

Bartlett, F. C. *Remembering*. Cambridge, England: Cambridge University Press, 1932.

Black, M. *The labyrinth of language*. New York: New American Library, 1968.

Blatt, S. J., & Quinlan, P. Punctual and procrastinating students: A study of temporal parameters. *Journal of Consulting Psychology*, 1967, *31*, 169–174.

Borow, H. Career development. *In* J. F. Adams (Ed.), *Understanding adolescence* (3rd ed.). Boston: Allyn and Bacon, 1976.

Bowlby, J. *Attachment and loss: Vol. 1. Attachment*. New York: Basic Books, 1969.

Bram, J. *Language and society*. New York: Random House, 1955.

Brenner, C. *An elementary textbook of psychoanalysis* (rev. ed.). Garden City, N.Y.: Doubleday (Anchor), 1974.

Bühler, C. *Der menschliche Lebenslauf als psychologisches Problem*. Leipzig: S. Hirzel, 1933.

Bühler, C. The course of human life as a psychological problem. *Human Development*, 1968, *11*, 184–200.

Carroll, J. B. *Language and thought*. Englewood Cliffs, N.J.: Prentice-Hall, 1964.

Cassirer, E. *An essay on man: An introduction to a philosophy of human culture*. Garden City, N.Y.: Doubleday (Anchor), 1944.

Chomsky, N. *Language and mind*. New York: Harcourt Brace Jovanovich, 1968.

Church, J. *Language and the discovery of reality*. New York: Random House (Vintage), 1966.

Conger, J. J. *Adolescence and youth*. New York: Harper and Row, 1973.

Cottle, T. J., & Klineberg, S. L. *The present of things future: Exploring of time in human experience*. New York: Macmillan–Free Press, 1974.

Cumming, E., & Henry, W. E. *Growing old: The process of disengagement*. New York: Basic Books, 1961.

Dardel, E. History and our times. *In* M. R. Stein, A. J. Vidich, & D. M. White (Eds.), *Identity and anxiety*. Glencoe, Ill.: Free Press, 1960.

Davitz, J., & Davitz, L. *Making it from 40 to 50*. New York: Random House, 1976.

de Grazia, S. *Of time, work and leisure*. Garden City, N.Y.: Doubleday, 1964.

Dickstein, L., & Blatt, S. J. Death concern, futurity, and anticipation. *Journal of Consulting Psychology*, 1966, *30*, 11–17.

Doob, L. W. *Patterning of time*. New Haven, Conn.: Yale University Press, 1971.

Douvan, E., & Adelson, J. *The adolescent experience*. New York: Wiley, 1966.

Eliade, M. *Cosmos and history: The myth of the eternal return*. New York: Harper and Row, 1959.

Elkind, D. Egocentrism in adolescence. *Child Development*, 1967, *38*, 1025–1034.

Elkind, D. *Children and adolescents: Interpretive essays on Jean Piaget* (2nd ed.). New York: Oxford University Press, 1974.

Epley, D., & Ricks, D. Foresight and hindsight in the TAT. *Journal of Projective Techniques*, 1963, *27*, 51–59.

Erikson, E. *Childhood and society* (2nd ed.). New York: Norton, 1963.

Erikson, E. *Insight and responsibility*. New York: Norton, 1964.

Erikson, E. *Identity: Youth and crisis*. New York: Norton, 1968.

Fenichel, O. *The psychoanalytic theory of neurosis*. New York: Norton, 1945.

Fraisse, P. *The psychology of time.* New York: Harper and Row, 1963.

Fraser, J. T. (Ed.). *The voices of time.* New York: Braziller, 1966.

Fraser, J. T. *Of time, passion, and knowledge.* New York: Braziller, 1975.

Freud, S. *New introductory lectures on psychoanalysis.* New York: Norton, 1933.

Freud, S. *Beyond the pleasure principle.* New York: Liveright, 1950. (Originally published, 1920.)

Freud, S. Character and anal eroticism. *In* S. Freud, *Collected papers.* Vol. 2. London: Hogarth Press, 1953a. (Originally published, 1908.)

Freud, S. A note upon the "mystic writing-pad." *In* S. Freud, *Collected papers.* Vol. 5. London: Hogarth Press, 1953b. (Originally published, 1925.)

Gallatin, J. E. *Adolescence and individuality.* New York: Harper and Row, 1975.

Gergen, K. J. *The concept of self.* New York: Holt, Rinehart and Winston, 1971.

Ginzberg, E. Toward a theory of occupational choice: A restatement. *Vocational Guidance Quarterly,* 1972, *20,* 169–176.

Ginzberg, E., Ginsburg, S. W., Axelrod, S., & Herman, J. L. *Occupational choice.* New York: Columbia University Press, 1951.

Gould, R. L. The phases of adult life: A study in developmental psychology. *American Journal of Psychiatry,* 1972, *129,* 521–531.

Gurwitsch, A. *Studies in phenomenology and psychology.* Evanston, Ill.: Northwestern University Press, 1966.

Guyau, J. M. *La genèse de l'idée de temps* (2nd ed.). Paris: Alcan, 1902.

Hall, C. S., & Lindzey, G. *Theories of personality* (2nd ed.). New York: Wiley, 1970.

Hall, E. T. *The silent language.* Garden City, N.Y.: Doubleday (Anchor), 1973.

Hartmann, H. *Essays on ego psychology.* New York: International Universities Press, 1964.

Havighurst, R. J. Youth in exploration and man emergent. *In* H. Borrow (Ed.), *Man in a world at work.* Boston: Houghton Mifflin, 1964.

Hunter, I. M. L. *Memory.* Baltimore, Md.: Penguin Books, 1964.

Jacobson, E. *The self and the object world:* New York: International Universities Press, 1964.

James, W. *Principles of psychology.* Vol. 1. New York: Holt, 1890.

James, W. *Psychology: Briefer course.* Cleveland: World Publishing Co., 1948. (Originally published, 1892.)

Jaspers, K. The axial age of human history. *In* M. R. Stein, A. J. Vidich, & D. M. White (Eds.), *Identity and anxiety.* Glencoe, Ill.: Free Press, 1960.

Kastenbaum, R. The dimensions of future time perspective, an experimental analysis. *Journal of General Psychology,* 1961, *65,* 203–218.

Kastenbaum, R. The foreshortened life perspective. *Geriatrics,* 1969, *24,* 126–133.

Kimmel, D. C. *Adulthood and aging.* New York: Wiley, 1974.

Kluckhohn, F. R. Dominant and variant value orientations. *In* C. Kluckhohn, H. A. Murray, & D. M. Schneider (Eds.), *Personality in nature, society, and culture* (2nd ed.). New York: Knopf, 1953.

Kluckhohn, F. R., & Strodtbeck, F. L. *Variations in value orientations.* New York: Harper and Row, 1961.

Köhler, W. *The mentality of apes.* New York: Vintage Books, 1959.

Korzybski, A. *Science and sanity.* Lancaster, Pa.: Science Press, 1933.

La Barre, W. *The human animal.* Chicago: University of Chicago Press, 1954.

Langer, J. *Theories of development.* New York: Holt, Rinehart and Winston, 1969.

Langer, S. K. *Philosophy in a new key: A study in the Symbolism of reason, rite, and art.* New York: New American Library, 1951.

Lee, D. *Freedom and culture.* Englewood Cliffs, N.J.: Prentice-Hall, 1959.

le Lionnais, F. *The Orion book of time.* New York: Orion, 1959.

Le Shan, L. L. Time orientation and social class. *Journal of Abnormal and Social Psychology,* 1952, *47,* 589–592.

Levinson, D. J., Darrow, C., Klein, E., Levinson M., & McKee, B. The psychosocial development of men in early adulthood and the mid-life transition. *In* M. Roff & D. F. Ricks (Eds.), *Life history research in psychopathology.* Vol. 3. Minneapolis: University of Minnesota Press, 1974.

Lindsay, P. H., & Norman, D. A. *Human information processing: An introduction to psychology.* New York: Academic Press, 1972.

Maas, H. S., and Kuypers, J. A. *From thirty to seventy: A forty-year longitudinal study of adult life styles and personality.* San Francisco: Jossey-Bass, 1974.

Maxwell, R. J. Anthropological perspectives. *In* H. Yaker, H. Osmond, & F. Check (Eds.), *The future of time: Man's temporal environment.* Garden City, N.Y.: Doubleday, 1971.

Monks, F. Future time perspective in adolescents. *Human Development,* 1968, *11,* 107–123.

Muuss, R. E. The philosophical and historical roots of theories of adolescence. *In* R. E. Muuss (Ed.), *Adolescent behavior and society* (2nd ed.). New York: Random House, 1975.

Nakamura, H. Time in Indian and Japanese thought. *In* J. T. Fraser (Ed.), *The voices of time.* New York: Braziller, 1966.

Needham, J. Time and knowledge in China and the West. *In* J. T. Fraser (Ed.), *The voices of time.* New York: Braziller, 1966.

Neisser, U. *Cognitive psychology.* New York: Appleton-Century-Crofts, 1967.

Neugarten, B. L. (Ed.). *Middle age and aging.* Chicago: University of Chicago Press, 1968.

Neugarten, B. L., and Associates. *Personality in middle and later life.* New York: Atherton Press, 1964.

Phillips, J. L. *The origins of intellect: Piaget's theory* (2nd ed.). San Francisco: Freeman, 1975.

Piaget, J. *The construction of reality in the child.* New York: Basic Books, 1954.

Piaget, J. Time perception in children. *In* J. T. Fraser (Ed.), *The voices of time.* New York: Braziller, 1966.

Piaget, J. *The child's conception of time.* New York: Ballantine Books, 1971a. (Originally published, 1927.)

Piaget, J. *Genetic epistemology.* New York: Norton, 1971b.

Pollio, H. R. *The psychology of symbolic activity.* Reading, Mass.: Addison-Wesley, 1974.

Posner, M. I. *Cognition: An introduction.* Glenview, Ill.: Scott, Foresman, 1973.

Rapaport, D. *Emotions and memory.* New York: International Universities Press, 1950.

Roff, M., & Ricks, D. F. (Eds.). *Life history research in psychopathology.* Vol. 3. Minneapolis: University of Minnesota Press, 1974.

Royce, J. R. (Ed.). *Psychology and the symbol.* New York: Random House, 1965.

Rozek, F., Wessman, A. E., & Gorman, B. S. Temporal span and delay of gratification as a function of age and cognitive development. *Journal of Genetic Psychology,* 1977, (in press).

Russell, J. L. Time in Christian thought. *In* J. T. Fraser (Ed.), *The voices of time.* New York: Braziller, 1966.

Schachtel, E. G. *Metamorphosis: On the development of affect, perception, attention, and memory.* New York: Basic Books, 1959.

Sheehy, G. *Passages: Predictable crises of adult life.* New York: Dutton, 1976.

Sorokin, P. A., & Merton, R. K. Social time: A methodological and functional analysis. *American Journal of Sociology,* 1937, *42,* 615–629.

Spitz, R. A. Ontogenesis, the proleptic function of emotion. *In* P. H. Knapp (Ed.), *Expression of the emotions in man.* New York: International Universities Press, 1963.

Straus, E. W. *Phenomenological psychology.* New York: Basic Books, 1966.

Stutterheim, C. F. P. Time in language and literature. *In* J. T. Fraser (Ed.), *The voices of time.* New York: Braziller, 1966.

Sullivan, H. S. *The interpersonal theory of psychiatry.* New York: Norton, 1953.

Super, D. E. *The psychology of careers.* New York: Harper and Row, 1957.

Toffler, A. *Future shock.* New York: Bantam Books, 1970.

Toulmin, S., & Goodfield, J. *The discovery of time.* New York: Harper and Row, 1965.

Uexküll, J. von. *Umwelt und Innenwelt der Tiere.* Berlin: Springer, 1921.

Uexküll, J. von. *Theoretische Biologie.* Berlin: Springer, 1928.

Wallace, M., & Rabin, A. I. Temporal experience. *Psychological Bulletin,* 1960, *57,* 213–236.

Wessman, A. E. Personality and the subjective experience of time. *Journal of Personality Assessment,* 1973, *37,* 103–114.

Whitrow, G. J. *The natural philosophy of time.* New York: Harper and Row, 1963.

Whitrow, G. J. *The nature of time.* Baltimore, Md.: Penguin, 1975.

Whorf, B. L. *Language, thought, and reality.* New York: Wiley, 1956.

Yates, S. Some aspects of time difficulties and their relation to music. *International Journal of Psychoanalysis,* 1938, *16,* 341–354.

Zern, D. The influence of certain child-rearing factors upon the development of a structured and salient sense of time. *Genetic Psychology Monographs,* 1970, *81,* 197–254.

Editors' Introduction

In our attempts to understand the objects and the processes of our world, we employ cognitive systems for symbolizing space, time, and quantity. Riegel forcefully argues that in our efforts to comprehend time, we have inadvertently used spatial and numerical concepts. For example, when we conceptualize quantity, we can progressively categorize things in nominal scales, order them by relative magnitudes with ordinal scales, and possibly measure them with equal-interval scales with arbitrary or absolute zero-points. When speaking of time, we perform operations parallel to those of quantification when we consider simultaneity, order, and relative and absolute durations. Piaget and other developmental researchers have shown that children's time concepts emerge in a continuum ranging from categorizations of simultaneity to formal operational constructions of "absolute" time. Riegel, however, contends that these concepts of time are highly limited in that they assess time from the implied single vantage point of a stationary, ideal observer—which is realistically unobtainable.

Riegel introduces the concept of *dialectical time,* in which time, like energy, is viewed as an "interphenomenon" produced by the interplay of multiple ongoing events. In dialectical time, multiple temporalities of many sequences interact to form a unique time that characterizes the event in its many aspects.

Riegel then demonstrates how a dialectical viewpoint can be employed to help us understand the temporality of such diverse phenomena as dialogues, historical change, memory, and individual life histories. For example, in reviewing a person's life history, we often tend to focus only on the changing person. However, the society is also changing in many ways that have direct bearing on the person's behavior. Riegel asks us to view individual development as involving the interaction of at least four dimensions; inner-biological, individual-psychological, cultural–sociological, and outer-physical. (See Albert and Jones's use of "templates" for a similar purpose.) An individual's life history (or, for that matter, any complex event) can then be viewed in terms of the interactions of events from each dimension.

Riegel's dialectical paradigm leads to a theory of time experience that can encompass personal subjective time, clock time, social time, and biological time within the same conceptual model.

2

Toward a Dialectical Interpretation of Time and Change

Klaus F. Riegel

In one of his insightful though complex treatments, Piaget (1970a,b) complained that unlike geometry for the study of space, there is no formal theory available for the study of time. By proposing the term *chronometry* for such a theory, Piaget seems to have assigned interval properties to it. Interval properties also characterize formal operational systems of Newtonian space and the concrete operational systems of projective space. In opposition to such an interpretation and in spite of all the measurements that we are taking, I argue here that our concept of time should realize a much simpler basis; that is, it should rely on categorical or relational systems. The need for arguing in this way at all indicates how much we have uncritically subordinated our time concept to the powerful geometry of space and how little we have succeeded in exploring the philosophical basis of time in an independent manner.[1]

[1] This chapter is one of my recent attempts to explore the concept of time and to apply it to the behavioral and social sciences. My earlier publications have dealt with the analysis of individual and cultural changes in developmental studies (1965), explanations and deterministic theories of development (1966, 1968), and qualitative growth models (1972). Some of these and several other papers on the psychology of development and history have recently been republished (1976c). My most recent explorations of the concept of time (1977) have been stimulated and strongly influenced by Donna Cohen's manuscript entitled "Time as Energy: On the Application of Modern Concepts of Time to Developmental Sciences."

Klaus F. Riegel ● Department of Psychology, University of Michigan, Ann Arbor, Michigan.

1. The Concept of Time

Relying on some major ideas from the theory of numbers and measurements (see Hölder, 1901; Stevens, 1951), the following analysis emphasizes the logical basis of time concepts. I begin with the simplest system of all, the categorical system, and then proceed to more complex realizations, to ordinal, interval, and absolute systems. Each of these systems relies on some additional axioms, which, in a hierarchical manner, always embrace those of the preceding levels. For the discussion of the concept of time, the four topics corresponding to the four measurement systems are simultaneity, direction, duration, and zero-point.

1.1. Simultaneity

What objects are in space, events are in time. During the very early stages in human life, there is no separation of stable objects and changing events. A child attends to objects only if they are in motion or if through the child's own motions (from gross movements of the body to minute movements of the lenses), the impression of such changes is created. Also for the mature person, the separation of stable objects and changing events remains arbitrary and artificial. Events involve objects, and objects change. Indeed, the connection is so intimate that only through the conceptual separation of space and time has it become possible at all to regard objects (being in space) as independent from events (being in time). If we wish to draw inferences about the concept of time at this early stage of the child's development, we would have to consider movements, actions, and events in themselves as representing such notions. As we shall see, however, movements, actions, and events are directional and therefore represent more appropriately the second stage in the development of the time concept.

Simultaneity is the most basic property of temporal descriptions, but it also has created some of the greatest difficulties. Coexistent objects (for example, the stars in the sky) define simultaneity as the intersection of world lines (that is, of the rays along which the light travels to the eye of the observer). Such a point has no temporal extension; time collapses upon a momentary, timeless instant. Simultaneity of this kind is not concerned with movements through time or changes of time. If such a pointilism were all that our analysis of time aimed for, the resulting description would be formulated in the spatial terms of classical natural science and philosophy, into which time enters only secondarily through comparisons of different timeless slates. Within these slates, all objects (or our perceptual impressions) retain their identity; everything is fixed and stable; nothing changes. For a fuller understanding of time, directional or relational properties have to be considered.

Simultaneity or coexistence in space depends on the notion of object permanence and seems easy to comprehend. It is comparable to the notion of equivalence or, more important, to that of identity in spatial descriptions and formal logic. Space itself—at least during the first stage of development and in the early periods of history—is characterized by the simultaneous existence and distribution of objects. For the concept of time, however, it is exceedingly difficult to determine whether or not two events are simultaneous, and this is not only a problem of measurement accuracy (When are two durations of performance exactly alike? When are two children of exactly the same age?) but has broader implications. For example, if one person is asleep and another person awakens him or her by producing a loud noise, is it reasonable to say that this interrupting event is simultaneous for both? Shouldn't we rather think of two time sequences that have their independent markings and are not strictly comparable in terms of any particular event, not to speak of their rate of progressions? The psychological recording of two events as simultaneous in time is so difficult to perform because it requires the simultaneous attending of two events by one person. If these events are different in kind and are experienced by different participants, simultaneity is even more difficult to explain.

Simultaneity provides temporal markings if the interaction of at least two event sequences is being considered, for example, the handling of the stop watch and the event observed. If we wish to determine, however, whether two performances extend exactly over the same duration or whether two children are of exactly the same age, we have to consider simultaneity at least twice, namely, the simultaneity of the start or birth and the simultaneity of the measurement here and now. In drawing inferences from such observations, we impose the ordered relation between start and measurement, and thus we have moved conceptually beyond the topic of simultaneity.

1.2. Direction

As important as, if not more important than, judgments of simultaneity are statements about movements and actions. Thus, it is not surprising that Piaget has linked the early apprehension of time to the experience of movements and the execution of actions. When Einstein raised the questions, "Is time immediate or derived? Is it integral with speed from the outset?" Piaget concluded rather readily that speed and movement are psychologically more basic than the abstract concept of time.

There is little reason to doubt his conclusion. When we try to recollect past events, we rarely reconstruct them with the aid of a temporal yardstick, such as the clock or the calendar. Various events appear spontaneously in our memory and provide the basis for temporal markings. These events do not appear as a haphazard collection, merely organized in terms of their simultaneity; there is also an apprehension of their movements and their temporal order. Moreover, such a relational organization does not occur along one single time dimension. On the contrary, we might apprehend a whole set of temporal orders, each representing a different time sequence, for example, when we reflect on our career in school, our vocation or profession, our family, our friends, or the political or economic situation. The interactions between these sequences provide temporal markings in the stream of events. Through these experiences, we also begin to apprehend the multidimensionality of the time matrix.

As immediate as the notion of order and direction seems to be to our understanding of time, just so difficult has it been to tackle this issue in science. Like the problem of simultaneity, these difficulties can be explained by the overriding traditional emphasis upon much more powerful time scales, that is, those of unidimensional, absolute time. The explorations of temporal order and direction are related to attempts in thermodynamics to reduce and explain time by a sequence of spatial conditions. Typically, these arguments are based on the example of two gases (substances) enclosed in two containers (space). If both containers were joined, the gases would slowly reach a mixed state. Because of the distribution of their particles in space, this state can be said to be temporally later than the state in which the two gases were less completely mixed or were separated altogether. The attempts to reduce time to a sequence of spatial conditions in the distribution of substances indicates a move away from the concept of absolute time in the classical natural sciences. Historically, it opened the way for the time concept in modern sciences, that is, in quantum physics and relativity theory.

1.3. Duration and Zero-Point

The notion of different event sequences crisscrossing each other but converging upon the momentary state of the reflecting observer is comparable to the notion of projective space during the concrete-operational period of late childhood. As described by Piaget and Inhelder (1956), this notion is soon replaced by the concept of absolute time, just as the concept of projective space is replaced by the concept of absolute space. As a prerequisite for achieving this transition, Piaget's notion of "decentering" attains significance. When the child succeeds in looking at the past (as well as into the future) from various angles or points of view, this transition has been achieved. A child who conceives of different time sequences as partially independent of each other and as interacting only at a few event points or knots has already abandoned the undifferentiated time concept of his or her former years and meets the standards necessary for the transition to a time concept implying the notions of duration and zero-point. More important, however, is the success of seeing the past (or the future) from the perspectives of different people, for the child to recognize, for example, that what appears to him or her as an intense period with many temporal markings is but an insignificant set of events to the teacher or that what appears as well timed to the child might appear as tardy to the parents.

In addition to the increased skill of decentering and thereby of viewing the past (or the future) from various perspectives, the transition to the stage of an absolute time concept is, ultimately, achieved through the notion of an ideal observer who is infinitely far removed from the present either into the past or into the future or, most likely, into both. This transition provides the interpretation that time is one-dimensional and uniform and stretches positively and negatively into infinity. Now, the temporal experience of the individual counts little and fills only an insignificantly small stretch of the universal time dimension. The individual takes no active part anymore but is subordinated to an impersonal time conception. In exchange, time has become quantifiable, although the procedures employed for achieving this goal remain rather arbitrary; that is, they rely on selected periodic systems, such as the solar year, the lunar month, or the swing of a standard pendulum.

1.4.   The Development and History of the Time Concept

The properties generally ascribed to time—that is, simultaneity, direction, duration, and zero-point—make it possible to construct a developmental sequence that coincides with what Piaget has described as stages in the child's development.

The developmental sequences of both space and time presuppose an apprehension of the concept of substance brought about by the differentiation of the subject–object relationship. The first stage leads to the apprehension of coexistent objects in space and of simultaneous events in time. However, space and time are not clearly separated, and in particular, the notion of coexistence applies more readily to organization in space than to organization in time. In the sense of ancient Greek philosophy, space is regarded as a void, that is, as that part not filled by coexistent objects.

The notion of direction and order leads to the apprehension of time as action and movement. Action and movement affect objects and create events. Although the separation of objects and events can never be distinct and clear, even a crude differentiation leads to the separate apprehension of space (through coexistent objects) and time (through consecutive events).

Further differentiation is achieved through generalizations leading to the apprehension of various interconnected event-sequences. This development is comparable to the acquisition of the concept of projective space. Whereas at earlier ages, different sequences were conceived of as independent and without either inner or outer connections, the older child begins to relate these sequences to one another and succeeds in viewing them from an increasing number of different perspectives. Especially when the degree of decentering achieved is high, the generality of the resulting concepts is extensive and, according to Piaget, provides a direct transition into the next stage of the absolute time concept of classical natural science and philosophy.

At the formal operational stage, the time concept has become an ideal abstraction. The individual observers have been eliminated from the systems; they have become subordinated to the uniform and universal time concept that they themselves have created during the course of history. Although now space and time are clearly distinguishable, they remain connected because they rely on the same form of mathematics and metrics. More precisely, the formalization of the space concept has been generalized to that of time. Time is regarded as a rather mysterious fourth dimension derived by analogy from the concept of space. Finally, with the development of thermodynamic theory, a reduction of the concept of time to that of space was promoted. This attempt, in turn, signaled a more radical break with the space and time concepts of classical natural sciences and philosophy.

As relational time is comparable to a single intrinsic event-sequence represented, for example, by a monophonic melody, absolute time is comparable to a single extrinsic standard, such as the meter and the bar lines on the score sheet. Both concepts are subordinated to a dialectical interpretation. Dialectical time is comparable to polyphonic music, in which various monophonic sequences are interwoven and in which temporal markings are generated by the harmonies and disharmonies of such a composition. Also, absolute time plays a role in such an arrangement—namely, as a selected monophonic sequence that serves as an extrinsic yardstick—and in particular, may aid in the synchronization of the different voices or instruments. In comparison to relational time, which is intrinsic, and in comparison to absolute time, which is extrinsic, dialectical time is both intrinsic and extrinsic, not one or the other.

2. Simultaneity, Observations, and Memory

2.1. World Lines

In the past, behavioral and social scientists have neglected the relationship between the time of the organism and the time of its inner and outer environment. In describing this relationship, Reichenbach (1958) and Reichenbach and Mathern (1959) adopted a definition from relativity theory. The age of an organism is characterized by its location on the world lines of the system. Development is the movement along the world lines. Chronological time is a functional unit of analysis for the world lines. Since it is external and arbitrarily measured by the clock on a linear scale, chronological time is not necessarily a meaningful index of development and change.

Since living systems are composites of various subsystems—for example, cells, cell clusters, organs, organ systems, or some other functional groupings—their analyses have to focus on their quantum, atomic, molecular, cellular, organismic, or behavioral states. Their development consists of series of processes along several of these dimensions with the time variable representing their transformations. For a demonstration of these transformations, we might construct the diagram of Figure 1, which represents liver cells posited on their respective world

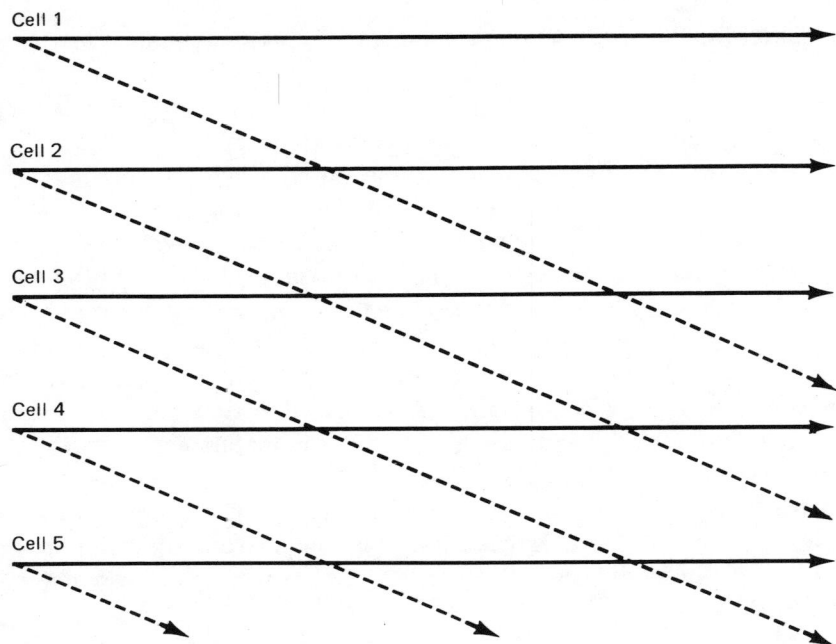

Figure 1.   Alternative world lines of liver cells.

lines. Each cell is dependent on others for the healthy development and functioning of the whole organ system, but there may be as many processes of development as there are cells (Birren, 1959).

Organ systems could be developing at different rates, determined, for example, by their metabolic conditions. The liver and the lungs may be changing rather fast, while the heart could be changing in a negative direction relative to the lungs or neuronal tissues. Thus, the world lines of Figure 1 do not necessarily remain parallel but could crisscross each other. World lines represent irreversible and universal orders, but a representation of this kind still leaves us with an arbitrary time scale for characterizing developmental processes within these interdependent functions. As shown by the dotted lines in Figure 1, it is conceivable, for example, to explore alternative directions for these world lines. However, in accepting sets of world lines for the representation of changes in a system, time loses its unidimensionality.

2.2. Measurements

Time is intimately connected with the observation of events. Since an observation is an irreversible process, we derive a sense of the direction of time from the perceptions of events. The time sense gives an element of organization to events, and we retain an equilibrium in our actions by regulating cues in time. With our actions, we create an entropy increase along biological world lines, but we are "neither the biological clock, nor a succession of time spans. . . . Man, the hierarchy of structures, lives in many structures and in many time coordinates. He does not live through one evolution, but emerges from many" (Meerloo, 1970, p. 132).

The most substantial experience with sidereal time consists in the recognition that a continuously greater number of events has occurred moment after moment. Such an outcome need not always be interpreted as a linear, horizontal accumulation of events but rather as a vertical transformation of events from microscopic to macroscopic levels. World lines at any level—molecular, cellular, organic, individual, or cosmic—are recorded as irreversible, but velocities may vary in the accumulation of different events. If we imagine an observer within a developing cell (as proposed by Leibniz's idea of the monads), the duration of physical processes in cell 2 might appear to be lengthened. But an observer in cell 2 would say the same about cell 1. The apparent time dilation is the result of the velocity perspectives. In special relativity theory, this topic has been called the *reciprocity of appearances*.

Relativity theory has introduced a multilevel conception of time and is concerned with the recordings of different clocks in different coordinate systems. Since it does not deal with the experience of the observer, it is not necessarily relevant to the behavioral and social sciences. Since relativity theory permits us to measure time in different coordinate systems, it has implications for an analysis of the organism in which we are trying to identify particular clocking mechanisms and their interactions. The problem thus centers again on the integration of the different levels of operations. We need to specify functions for several levels and need to relate the events on these different levels of analysis to each other. Perhaps we need to develop a construct of time that cannot be described in common experiential terms. Time may be explained as an energy, an interphenomenon, that cannot be perceived directly, and yet its basic functions may be associated with observable events.

The measurement of sidereal time permits us to order the events in the form of a series, E_1, E_2, E_3, . . ., E_n. But if we impose such a horizontal order on the events by the use of a clock, the nature of the event order remains insufficiently explored. In Popper's (1958) terms, the observer would have become the causal center in this analysis. If we reject such a time concept and consider a more complex construct, we obtain the vertical T_2 axis of Figure 2 rather than a point in single series of events. This two-dimensional time model has been introduced by Dobbs (1965) and has been modified here to explain the energetic property of time-space.

In Figure 2, an instant of "now" is represented by the intersection of times at some point, P; all Ps form a "now line." With an added dimension of time, an infinite number of alternative new points can be realized, a possibility not permitted in a one-dimensional model. As also illustrated in Figure 2, a change in velocity of the event accumulation could result in curvilinear now lines. The vertical axis can assume many functional forms. Consider, for example, a leaky faucet with drops of water falling rhythmically from the opening. The time rhythm results from the interaction of two processes, gravity and surface tension. The interaction of events on two dimensions creates a large set of time–energy functions.

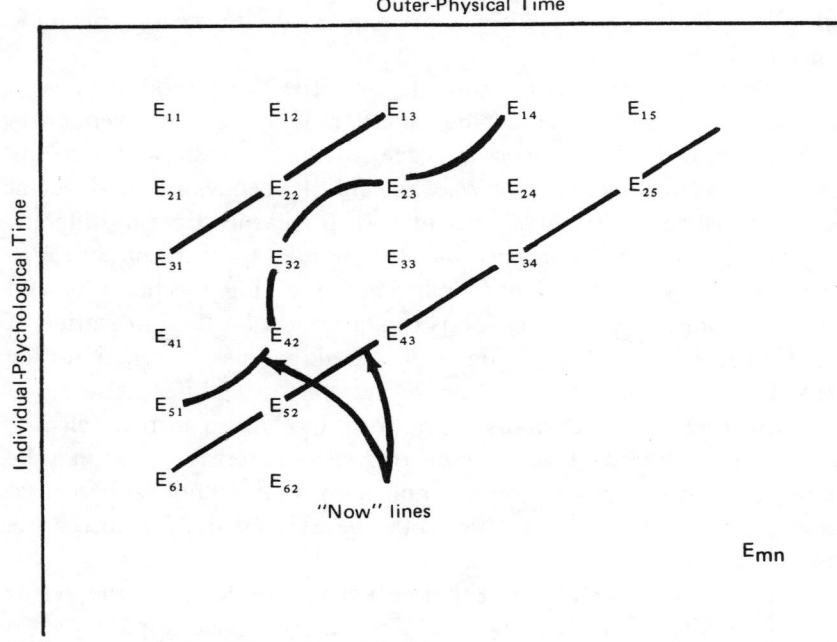

Figure 2.   Two-dimensional time matrix with world lines of events accumulating along several levels of analysis and with alternative "now" lines.

Energy transformations in metabolic processes may provide an adequate definition of time-dependent changes in the organism. According to Hoagland (1933, 1936), time judgments are dependent on the velocity of oxidative metabolism in parts of the brain; they are determined by the chemical velocities of biological reactions. EEG alpha rhythm changes in frequency when the internal temperature is modified by diathermia. Raising the internal body temperature increases the speed of reaction so that the psychological clock runs faster than a laboratory clock: about 2 minutes of subjective time equals 1 minute of clock time. Lowering the body temperature makes clock time seem faster; about 1 minute of subjective time equals 2 minutes of clock time. Supportive data were obtained when an individual tapped and counted seconds at different induced body temperatures.

2.3. Memory Processes

James, in opposition to the time atomism prevailing in the British sensualism of Locke, Hume, and Baine, and in opposition to the French rationalism of Descartes, emphasized the stream of experience that knows almost no boundaries and markings. Husserl, in comparison, gave detailed attention to momentary synchronic states as well as to their temporal blending in retrospection and prospection. The schema recently advocated by Kvale (1974) combines these interpretations with the views held by Merleau-Ponty (1962).

As shown in Figure 3, the horizontal line represents the world lines of a "retentionalizing" individual with three event markings, A, B, and C. The vertical lines at A and B represent some other world lines, indicating the changing state of the memorized events. The process of recollection at the time of events B and C is indicated by obliqued "now-lines." Thus, at the time of event C, the previous B appears in retention as B′, and the previous A appears as A″. From C, A″ is now seen through the intermediate B′. As stated by Husserl (1964): "Retention itself is not an 'act' but a momentary consciousness of the phase which has expired and, at the same time, a foundation for the retentional consciousness of the next phase. Since each phase is retentionally cognizant of the preceding one, it encloses in itself, in a chain of mediate intentions, the entire series of retentions which have expired" (pp. 161–162).

Figure 3 shows a two-dimensional time plot with both dimensions representing changes in the individual. The horizontal dimension

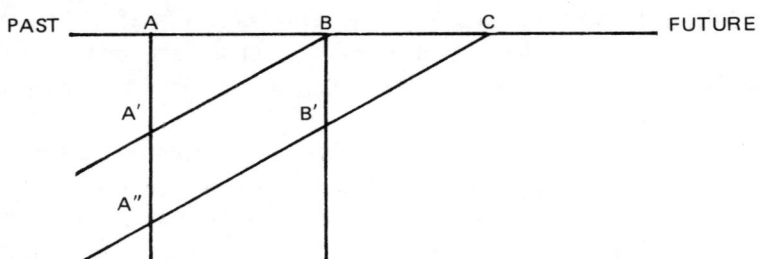

Figure 3.   Retentionalizing of impressions (from Kvale, 1974, Figure 1, p. 16).

depicts the remembering individual, who, in interaction with other individuals and social conditions, brings about the occasions B and C at which he or she retrospects about past events. The vertical dimension represents the experienced events themselves as they are changed in interaction with other events either experienced prior to them or after the criterial events have occurred. The obliqued lines are "now lines" that represent cross-sections through the memory at particular moments in time. Underlying this representation is the notion that memory, as well as the human being and society, is in a constant state of change and that because of these changes, earlier states can never be attained again. With any new experience, the structure of the memory becomes increasingly complex; with every new event added onto the horizontal dimension, the slanted "now lines" cover a wider and wider range of past events.

Activities are determined not only in interactions with co-occurring events and through retentional interactions with events of the past but also by the actor's intention. Kvale (1974) continued:

> The unity of the present with the past is thus constituted by retentions, and at the forward end of the perceptual arc the protentions join the present to the future. When listening to an enduring tone I more or less implicitly expect it to continue, to change, or to stop; this expecting consciousness is an integral part of the consciousness of the tone. In its most basic form, a protention is an empty intention with an open indeterminateness, constituting the formal basis for eventual more explicit contentual expectations of what is coming. (p. 16)

The approach advocated by Kvale is diametrically opposed to the efforts by most experimental and developmental psychologists. Both have preferred to study abstract performances, attitudes, or behaviors rather than retrospective memories and the prospective intention of individuals; thus, they have failed to recognize that most of these objectified products are experientially empty for the individual. They have disregarded that individuals care little about their performances unless they are understood within the context of their experienced past.

Although much more systematic than individuals who are casually searching their memories for events from the past, historians have always relied on recollections of such information. Historians utilize whatever means available—for example, archives, books, treatises, documents, and advice by other experts—in order to reconstruct the past as accurately and comprehensively as possible. Nevertheless, their attempts are hampered because the events to be reconstructed have rarely been experienced by the historians themselves but are made known to them through the mediation of other individuals. Therefore, the authenticity of the source material may be threatened to the extent to which the abstractness of psychological variables distorts the meaningfulness of the psychological interpretation.

2.4. Implications

The analysis of time has to consider an array of world lines: atomic, molecular, cellular, organismic, individual, cultural, cosmic. Traditionally, development was thought to occur only at the higher levels; atoms were usually regarded as "lifeless" and nondeveloping. Not only is the implied distinction between living and nonliving substances questionable on theoretical grounds, but its analysis also fails to take account of interactions across world lines that take on the attribution of transforming energy.

The concept of time as energy was derived from modern physics. The energetic property of time was demonstrated for behavioral and social sciences in a discussion of the temperature(energy)-dependence of temporal judgments. In extension, the concepts of individual-psychological and cultural-sociological time are introduced. Through their activities, individuals change historical conditions, which in turn change the individuals. Historical time, as traditionally conceived, is restricted to sequences of rigidified products; historical time as energy represents transformational activities, history itself.

3. Event Sequences, Dialogues, and Synchronization

The most fundamental problem of written or spoken language consists in the transformation of sets of coexisting impressions or ideas into the sequential order of linguistic expressions. This problem is not narrowly restricted to language and to the common contrast between semantic structure and syntactic organization but touches upon the simultaneity and sequential order of thought, upon harmonic and melodic structure in music, upon simultaneous-logical and the developmental-dialectical analysis in the sciences, and, basically, upon the distinction between the concepts of space and time.

3.1. Narrations[2]

The temporal structure of narratives has recently been analyzed by Labov and Waletzky (1967). With a similarity to Wundt's (1911–1912) interpretation, the task of narration is regarded as the transformation of a set of instantaneous states of objects and events into a structure sequentially ordered within itself as well as transformationally related to the set of conditions that it aims to describe. Thus, a narrative coordinates a series of nonlinguistic events with a sequence of linguistic expressions. Frequently, the latter strengthens the structure of the event sequence, which by itself might be unintelligible in its temporal organization to an outside observer.

The problem encountered is a familiar one to the writer and the historian. In particular, Carlyle is credited with handling these difficulties in a most eloquent manner. As stated by Clive (1969):

> In ordinary narrative history, "A" occurs; then "B" occurs (possibly, but not necessarily, caused by "A"); then "C" happens, etc. This of course, is a false rendering of actuality, since not only do many other events transpire simultaneously, along with "A," "B," and "C"; but it is also true that "A," "B," and "C," like all historical events, are anchored in the past and have repercussions in the future. The historians' usual means of dealing with this basic problem is to make use of phrases such as "meanwhile," "at the same time," "while this was happening here, that was happening there"; and to spell out, in so many words, both the background and the aftermath of the events he is narrating.
>
> The trouble with these stylistic devices is that they completely fail to capture the historical *process,* in which nothing is stationary, and everything is constantly in motion and in flux, in which growth and decay proceed at the same time; in which events do not occur in isolation, but are related to each other in a constantly shifting network of interconnections. What

---

[2] More comprehensive analyses of narratives, and especially of dialogues, by this author have appeared elsewhere (1976b). Other contributions to this topic have been made by Freedle (1975, 1977), Gunter (1974), Harris (1975), Volosinov (1973), Testa (1970), and several others mentioned in these publications.

Carlyle does is not to evade this problem, but to face it head on, by using linguistic and literary devices to create the equivalent of living reality. (p. xxxvi)

Narratives are produced only where there are listeners to listen. During the narration, these listeners remain relatively passive and rarely interact with the producer of the story. If both the narrator and the listener become active participants, the narration shifts into a dialogic exchange. Now, either of the two might tell the other about a different series of event states experienced, or both might refer to the same series of events, which they elaborate in form of alternative interpretations. Thus, a dialogue represents not only a sequential coordination of one or two series of event states with the narrations of one or two speakers but also deals with the coordination of the two speakers' performances with one another.

3.2. Situational Dialogues

A dialogue can be represented by a chain of arrows linking the utterances of speakers A and B in their temporal order. In comparison to narratives or *monologues,* there is a systematic alternation between the two participants in such a *dialogic chain.* The complexity of both types of sequences increases if some connections are going further back than to the immediately preceding utterances, that is, from A_3 to A_1 and not only to A_2 in the monologue, and from A_3 to A_2 and not only to B_2 in the dialogic chain shown below:

Dialogic Chain
$$A_1 \longrightarrow B_1 \longrightarrow A_2 \longrightarrow B_2 \longrightarrow A_3 \longrightarrow B_3 \ldots$$

Monologue or Narrative
$$A_1 \longrightarrow A_2 \longrightarrow A_3 \longrightarrow A_4 \longrightarrow A_5 \ldots$$

In the simplest form of a dialogue, each of the two speakers always relates his or her new statement to both the last statement made by his or her opponent and to that made by the speaker himself or herself. In such *simple dialogues,* each statement is connected with the last two preceding ones. In *complex dialogues,* connections span across more than two preceding utterances, one made by A, the other by B. Each statement must be consistent with the proponent's own previously expressed views and must represent an equally consistent or systematically modified reaction to all earlier statements by the other participant in the dialogue.

Simple Dialogue

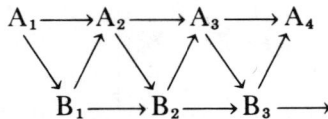

Complex Dialogue

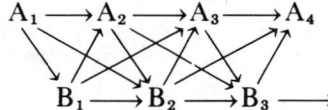

If a reflective coordination, as in simple or complex dialogues, did not take place, the exchange would degenerate into alternating monologues, in which each speaker would merely follow up on his or her own earlier statements without reacting to the other speaker's elaborations. The other speaker's statements would thus become distractive interruptions and the only remaining dialogical feature of such a performance would consist in the alternations between the participants. If these alternations cease, too, we approach a condition that Piaget (1926) has described as collective monologues. As in the tuning of the instruments before the music begins, two or more speakers continue their productions uninterruptedly and parallel to one another.

In contrast to monologic chains, dialogues are composed of triangular subsections, which I will call *dialogical units.* Such units indicate the dialectical character of the communication process. The original statement represents a *thesis,* which is denied, challenged, or modified by the other speaker. The second speaker's statement, therefore, represents an *antithesis.* Even a simple confirmation by a nod of the head is a challenge to the first speaker, who, in a second statement, takes notice of it and integrates this message in the form of a *synthesis.* But as this synthesis is uttered, the second speaker may propose a deviant interpretation, which, from his or her point of view, integrates his or her own former statement (thesis) with the statement of the opponent (antithesis). The contradictory relationship between any two statements makes the empirical description of dialogues rather difficult. Traditionally, we are used to categorizing observations by their similarity and not by their divergence. For the analysis of the dialogue, we have to do both.

The dialectical process that constitutes a dialogical unit indicates the reflective character of all dialogical interactions. An antithesis is not merely a reaction to but reflects upon the thesis backward in time. The thesis provokes and anticipates the antithesis; the antithesis modifies and reinterprets the thesis. None can be thought without the other. And what holds for the thesis and antithesis also holds for their relationships to the synthesis. Thus, each utterance in the dialogical exchange represents a thesis, an antithesis, and a synthesis at one and the same time—depending on which of the triangular subsections we focus upon. All adjacent statements are interdependent. Consequently, all arrows should have heads pointing in both directions. The unidirected arrows used in the diagrams are simplifications signifying the flow of physical time in the dialogue.

A statement is of no interest at all and bare of any meaning if it never becomes a part of some outer or inner dialogue. It becomes significant only if it is incorporated into a dialogical temporal structure, regardless of whether such a structure emerges immediately or many months later. A statement in complete isolation is like the sounds in the forest occurring in the absence of any listener. We might question with Wittgenstein their existence, but as we begin to think about these sounds, they enter into an inner dialogue and, therefore, gain meaning regardless of whether they "really" exist or not. All reality lies in the dialogical, or rather in the dialectical, process.

In a successful dialogue, each speaker assimilates the other person's statements and accommodates his or her own products so that they elaborate the preceding viewpoints. If this were not the case, the dialogue would either degenerate into alternating collective monologues or would converge into a repetitive cycle in which each speaker merely repeats or reaffirms what has been said before. Such repetitive cycles would be indicated in the diagrams by any triangle with identical numerical subscripts. Like the subsections in the truss of a bridge, repetitive or recursive operations provide stability to the dialogues. But dialogical exchanges can not subsist on recursive productions alone. The melody has to be varied and the cycle has to be broken by contrastive operations (Riegel, 1974). In this way, the topic is either moved into new divergent directions or converges upon—is integrated with—previous arguments.

3.3. Developmental Dialogues

The development of dialogues depends on the temporal coordination or synchronization of the mother's and the child's experiences and operations. Just a few weeks after birth, the child's and the mother's actions have already become finely tuned to one another. The child begins to look at the mother's face; when the mother moves, the child follows her with his or her eyes. When she speaks, the child might look at her mouth; when she stops, the child might vocalize and switch his or her attention from her mouth to her eyes (Lewis & Freedle, 1972). The temporal coordination of activities by the mother and her child represents the basis for the development of dialogues at more mature sociolinguistic levels.

Early in life, the mother has to be highly restrictive in her communication with the child. While she can rely on a large repertoire of signs, rules, topics, and roles, the child has virtually none of these acquired forms of operation available. Subsequently, the mother engages in directed efforts of giving instructions to the child or of naming objects. But aside from her conscious efforts, she talks and sings along, and the child eventually follows her activities and participates in them. During these weeks and months, a rather rapid expansion of the child's repertoire—and thus of the sign system shared between mother and child—takes place. During the following years, the "primitive" dialogue is transformed into a "scientific" dialogue (Lawler, 1975). Both speakers are able to select topics, disregard information, and engage in ploys (Harris, 1975).

The communication system shared by the mother and her child is at first private to both of them. As the signs become more and more congruent with those of the natural language, the mother functions as an intermediary between her child and society. She mediates between the cultural-sociological dimension of society and the individual-psychological dimension of her child. Thus, the developmental dialogue is more than an exchange between two individuals; society and history, through the mother, participate in it. The same conditions prevail in a situational or short-term dialogue. Each of the two speakers represents a particular cultural-sociological group, with its particular background, preferences, and goals. Communication between the participants is possible only because both speakers and both groups that they represent are, in turn, part of a larger cultural-sociological system and thus share a common and fundamental basis for coordination.

Similar arguments have been made by Marxist structuralists for the more remote interactions between authors and their readers (see Fieguth, 1973; Mao Tse-tung, 1968; Riegel, 1976b; Schmid, 1973). The author's task consists in transmitting to the reader the knowledge and the direction generated in the society throughout its history; this task is accomplished through the author's participation in these cultural-sociological efforts. Thus, the author, like the mother, functions as an intermediary between the knowledge- and direction-seeking individuals and the values and ideas of the society; he or she transforms the values and ideas of the society in order that they may be better understood by the individuals of the contemporary generation.

The coordination and synchronization of their efforts with those of the reader and the child, respectively, is most important in the author's and in the mother's task. Neither the reader nor the child should be overburdened or underburdened. Information has to be given at the right moment, in the right amount, and of the right kind. This effect is achieved through continuous developmental interactions with the information seeker. The author may become too abstract and remote; he or she may progress too fast or lag behind. The mother, in a more concrete sense, has to speak as she influences her child, but she also has to listen and change her own activities in accordance with the development of her child. In both cases, the synchronization of two event sequences—for example, the development of the child and the development of the mother—is of utmost importance. The process of synchronization is the most important issue in temporal description.

3.4. Synchronization of Event Sequences

What holds for the dialogical interactions between mother and child holds also for any interactions between any two individuals, for example, for the changing interactions in the career development of husband and wife.

As summarized in Table 1, an individual, after leaving school or college, will enter his or her first occupational career. While the attainment of an occupational role is primarily dependent on social condi-

Table 1. Levels and Events in Adult Life

Level (yrs.)	Gradual Changes				Sudden Changes
	Males		Females		
	Psychosocial	Biophysical	Psychosocial	Biophysical	
I (20–25)	College/first job Marriage First child		First job/college Marriage	First child	
II (25–30)	Second job Other children Children in preschool		Loss of job	Other children	
III (30–35)	Move Promotion Children in school		Children in preschool Move Without job Children in school		
IV (35–50)	Second home Promotion Departure of children		Second home Second career Departure of children		
V (50–65)	Unemployment Isolation Grandfather Head of kin	Incapacitation	Unemployment Grandmother Head of kin	Menopause	Loss of job Loss of parents Loss of friends Illness
VI (65+)	Deprivation	Sensorimotor deficiencies		Widowhood Incapacitation	Retirement Loss of partner Death

tions, which, in turn, are the reflection of cultural standards at a particular historical time, service in the army is biologically determined insofar as in most societies it applies to males only. The marital role requires sexual maturity of both partners. But in all these instances the separation of biological from psychological and sociological determinants is difficult to achieve.

At the second developmental level, determining events are the birth of other children coupled with a complete or partial loss of the wife's employment, a change in the job or promotion of the husband. A child cannot be born unless a social or at least biological marriage has taken place; a job cannot be changed or lost unless it has been held before. As obvious as these conclusions are, they need to be emphasized because they yield the most important temporal markings of the adult's life. They play an important role both for the recollection of minor events at a later date—such as an accident, the purchase of a household item, a birthday or party—and for the temporal markings of what individuals might perceive as crises in their lives.

At the third developmental level, the determining events include the execution of specific parental roles during the children's preschool years, especially by the mother; changes in job and promotion of the father; a move to a larger house or to a different location, etc. A clear delineation of a developmental level corresponding to and determined by the children's school years is possible only for families with a few children, narrowly spaced by birth over a short time period. This limitation indicates the cultural-sociological dependence of adult life periods. Most of them can be identified only for members of small nuclear families in industrialized settings. Agricultural societies with elaborated kinship traditions do not allow for subdivision of the course of life in distinct periods founded on other than biological determinants. Large kin groups experience closely spaced arrivals of children that do not allow for the sectioning of the life span by generational shifts.

During the fourth developmental level, the mother may prepare for or begin her second career. Few changes except those of promotion or shift in assignments may be experienced by her husband. However, the departure of the children from home in preparation for or pursuit of their own adult development and careers affects their parents profoundly. If these departures are accompanied by—what is becoming increasingly likely at the fifth level—the death of several of the grandparents, the status of the parents is even more drastically altered.

Both husband and wife may now attain the top position among the members of the extended family, with their own children ready to marry and grandchildren to be expected.

At the sixth developmental level, individuals become increasingly vulnerable to dismissal, unemployment, and disease. The death not only of the parents but of the partner, friends, and relatives may create personal crises; these incidents are likely to occur with greater frequency. While most of these events are brought about by uncontrollable outer-physical circumstances or are due to unavoidable inner-biological changes in the adult, one of the most decisive final affronts, mandatory retirement, is caused by conventional regulations. Retirement provides one of the last insults to adult persons and initiates their progressive social deterioration.

Undoubtedly, the presentation of the life course sectioned into temporal levels or periods reflects the biases of the traditional Western society. It also distorts the particular interpretation of time that I am promoting, that is, an interpretation emphasizing the conflicts and synchronization of event sequences that are constituted by temporal markings that in prospect or retrospect might be recognized as "objective" dates. For example, if at some later occasion, one of the partners were to recollect some episodes of the past, he or she would most likely focus upon events such as the birth of a child, the move to a different house, or the loss of a job. For temporal markers to be assigned to these episodes, the significance of the events has been brought about by their interactive implications, for example, by their effect upon the other children, by the attitudes of the old and new neighbors, and by the role of the employer and of other employees. As these interactions are recognized, a more distinct temporal order of the events becomes apparent. Only as a very last step are external, objective dates assigned to the sequence of interactive events.

3.5. Implications

The careers of common men and women often represent deficient forms of structured time sequences. The major events affecting individuals are arbitrarily imposed upon them by social and legal regulations, for example, their departure from school, recruitment into military service, job appointments and dismissals, and, ultimately, retirement. Other changes are brought about by cultural-sociological or outer-physical catastrophes, for example, depressions, inflations, revolutions, wars, droughts, floods, fires, and earthquakes. Only the inner-biological determinants seem to follow a predictable order, first revealing the individual's maturation, the birth of children, and later, increasing proneness to incapacitation, illness, and death.

Among the few favored individuals, the life span of an academic scientist allows for greater structural variations and for more sensible structural transformations during his or her developmental progression. Since also the structure of the scientific community and its historical changes have been explored in greater detail than those of most other groups, I devote here a good deal of attention to the interactive development of a scientist and his scientific discipline. This will be done after I have described some systematic attempts to analyze the interdependence of individual-psychological and cultural-sociological changes.

4. Development, History, and Science

In the preceding two sections, the synchronization of two individual developmental sequences were discussed: the dialogical development of mother and child and the coordination of the career development of husband and wife. In the following two sections, the interdependence of individual development and cultural history is discussed, first in the systematic form of developmental research designs and second by an example of interpretation from the history of the behavioral sciences.

4.1. The Analysis of Individual and Cultural Changes

As shown by Schaie (1965) and Baltes (1968), certain variables—for instance, the amount of mobility—may yield developmental gradients increasing in magnitude from generation to generation. If these increases are linearly related to age and if, furthermore, we were to assess age differences by the traditional cross-sectional method—that is, by testing at one time samples from different age groups (which thus represent different generations or cohorts)—results might be obtained like those indicated by the heavy line in Figure 4. Curves like the one in Figure 4 are very familiar to developmental psychologists but represent mere artifacts because neither the generation nor the time-of-testing effects have been controlled as contributing factors.

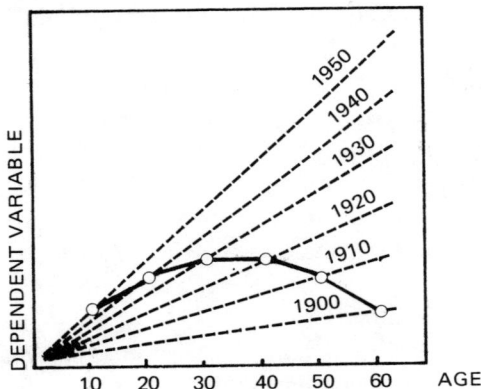

Figure 4.   A hypothetical example of the effects of generation differences on the results of a cross-sectional study. Broken line, generations; solid line, cross-section 1960 (from Baltes, 1968, Figure 1, p. 12).

The proper analysis can be best explained in reference to Table 2, which lists the age of three cohorts (born around 1850, 1900, 1950) over two times of testing (1920 and 1970). Comparisons within the rows of the table represent longitudinal designs; those along the diagonals from the lower-left to the upper-right cells represent cross-sectional designs; a third basic design embedded in the data, the time-lag design, has never been discussed by developmental psychologists. It compares cohort differences at various times of testing within specific age groups, that is, within the two columns of Table 2. As in the examples used before, Table 2 represents a two-dimensional time plot in which one dimension

Table 2. Time of Cross-Sectional Testing (Now Lines) in Developmental Research Designs

	Age	
Cohort	20	70
1850	1870	1920
1900	1920	1970
1950	1970	2020

indicates the individual-psychological age and the other cultural-sociological time. Cross-sections taken, for example, in 1920 and 1970 represent now lines.

None of the three basic developmental designs measures in an unconfounded manner either age, cohort, or historical time (time-of-testing) differences. An inspection of Table 2 shows that results from cross-sectional designs (CSD) confound age (AD) and cohort differences (CD); those from longitudinal designs (LOD) confound age (AD) and historical time differences (TD); those from time-lag designs (TLD) confound historical time (TD) and cohort differences (CD).

According to Baltes (1968, p. 156), these conclusions can be summarized as follows:

$$CSD = AD + CD$$
$$LOD = AD + TD$$
$$TLD = TD + CD$$

If we solve these equations for any one of the three right-hand terms, we obtain the following results:

$$AD = 1/2 \ (CSD - TLD + LOD)$$
$$CD = 1/2 \ (TLD - LOD + CSD)$$
$$TD = 1/2 \ (LOD - CSD + TLD)$$

Thus, it is in principle possible to obtain estimates of the "pure" effects of age, cohort, or historical time differences, but such attempts have always to rely on the joint utilization of all three basic designs: cross-sectional, longitudinal, and time-lag. This conclusion has far-reaching implications for the scientific disciplines involved and emphasizes the interactive basis for gaining knowledge, such as the analysis of temporal sequences. Psychology may justifiably describe individual-developmental changes, sociology may describe cohort or generational differences, and history may describe changes with chronological time. But these disciplines should not remain in isolation. Their contributions have to be recognized in complementary determination. Each alone produces abstract results and fictitious interpretations.

4.2. Life Histories

In the preceding discussion, I demonstrated the interdetermination of individual and cultural changes in a systematic manner. In the following discussion, I rely on the idiographic description (Riegel, 1975a) of the life history of a well-known psychologist in order to demonstrate for this particular, though general, case the interdependence of individual-developmental and cultural-historical progression.

The reliance on famous persons is useful for our purpose because their life histories are sufficiently well documented and their temporal structure has been marked, for example, by the records of their publications. Also, their impact upon the scientific community and the scientific community's impact upon them can be fairly well assessed. In a simplified manner, we might select, therefore, the major publications of a well-known scientist, his appointments at various locations, and the ranks attained and list them as major markings of his academic career. In realizing that these events are the products of the temporal interactions of the individual with his social community, we derive the following synopsis of the academic life of one famous psychologist, Wilhelm Wundt.

Wundt was born in 1832 and entered the University of Tübingen in 1851, where he stayed only for one year before he moved for four years to Heidelberg. Part of 1856 was spent in Berlin under the influence of Mueller, Magnus, and du Bois-Reymond.

The next decade, 1856–1865, has been described as "presystematic" by Boring (1957). Nevertheless, Wundt laid the foundation for most of his later work during this period. Between 1857 and 1864, he was Dozent (assistant professor) in Heidelberg and, presumably, under the influence of Helmholtz. In 1858, Wundt published the first section of his book, entitled *Beiträge zur Theorie der Sinneswahrnemungen,* which was completed in 1862 and has been regarded by Titchener as a blueprint of his lifetime work. In 1863, he also published *Vorlesungen über die Menschen und Thierseele,* which includes a brief outline of his *Völkerpsychologie,* a topic and program of investigation that was to lay dormant until the beginning of the 20th century, when Wundt devoted much attention to it.

Between 1865 and 1880, Wundt revealed his first paradigmatic orientation. He presented his work on perceptual and motor processes, with an emphasis on sensory elements and their compounding into complex units. Although much broader and more comprehensive than later critics have made us believe, Wundt was never able to shake off the impressions that he created during these years, the impressions of promoting a psychophysiological elementalism. This paradigmatic orientation is most distinctly represented by his two-volume work entitled *Grundzüge der physiologischen Psychologie,* which appeared between

1873 and 1874. He also became *Ausserordentlicher Professor* (associate professor). In 1874, he moved for one year to Zurich and 1875 to Leipzig, where he opened his psychological laboratory in 1879.

The decade between 1880 and 1890 was almost exclusively devoted to philosophical writings. In 1881, he started the journal *Philosophische Studien* (although it included predominantly psychological reports and was later renamed in this sense). Between 1880 and 1883, he published two volumes on logic, in 1886 one on ethics, and in 1889 a book entitled *System der Philosophie*. He concluded the decade by serving as the Rector of the University of Leipzig.

The 1880s can be regarded as a preparatory period for promoting his second paradigmatic orientation representing his three-dimensional theory of feelings, first elaborated in his *Grundriss der Psychologie* of 1896. In contrast to the earlier psychophysiological and psychophysical theorizing, Wundt tried systematically to establish psychology on a more independent basis with less reference to physiology and physics. Although the extensive research on this topic created a considerable stir during Wundt's time, later psychologists paid little or no attention to it. Wundt remained branded as an introspective elementalist.

A similar disregard was shown for his extensive writings in "ethnopsychology," during the first two decades of the 20th century until his death in 1920. This approach—as already outlined in his "Vorlesungen" of 1863—was to supplement the experimental methods promoted by the two earlier paradigmatic orientations. By an analysis of the customs, habits, and languages of society as objectifications of the human mind, a second route for the study of psychology was to be provided. Experimentation is analyzing the mind from within; ethnopsychology analyzes the mind (more precisely, its products) from without. Each approach supplements the other. But in spite of the significance that Wundt assigned to it and in spite of the extensive debates and controversies that ethnopsychology created among anthropologists, linguists, and historians, present-day psychologists are hardly aware of Wundt's two-pronged attempt at a comprehensive integration of the behavioral and social sciences.

4.3. The Interaction of Individual and Cultural Event Sequences

In Figure 5, the changing interactions between the life of an individual and the history of a scientific community are presented orthogonally to each other. Again, we consider the life of Wilhelm Wundt.

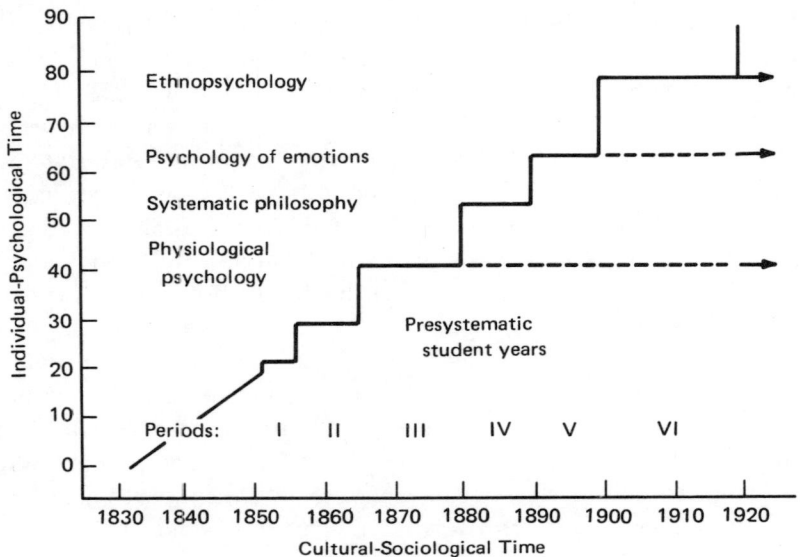

Figure 5. Two-dimensional time plot of the changing individual-psychological roles of one eminent scientist, Wilhelm Wundt, participating in three paradigmatic orientations at different cultural-sociological times.

The first period (1851–1856) of his academic life found Wundt among other students under the influence of his teachers. During the second period (1856–1865), he gained more independence but was likely to remain under the shadow of Helmholtz. Wundt revealed his first paradigmatic orientation during the third period, from 1865 to 1880. During this time, he became an influential teacher and writer. The stigma of this period, that of sensory elementalism, was not to be removed from him throughout the rest of his life and not even today. Being occupied with the administration of his laboratory, with academic duties, and with the publication of several philosophical works, the fourth period (1880–1890) must have led to a change in his academic role. Instead of directly interacting with his students he is likely to have been hidden behind his assistants, doctoral candidates, and visitors. Although addressing his collaborators in small seminars, he reached his students only through his large lectures and his many publications.

The fourth period has been regarded as a preparation for his second paradigmatic orientation, represented by his research on and theory of feelings during the fifth period (1890–1900). Because of the disruption of his interaction with a new generation of students, and in spite of the intensive debate among his followers and opponents, his theory never became an accepted part of psychological inquiry. The same fate befell his extensive efforts during the sixth and last period (1900–1920) to develop a third paradigmatic orientation, representing ethnopsychology. Although this work was heatedly debated in areas other than psychology, not even Wundt's associates and former students seem to have paid much attention to it. During his later years, Wundt must have exerted little direct influence upon the new generation of potential converts. He was retired, although he remained committed to scientific and editorial duties as well as to his extensive writings.

Wundt's career shows rather distinctly the changing interactions between the individual and the cultural group. The markings of the individual's life cannot be understood unless the changing social conditions are considered. The same holds true in the opposite direction: the changing status of the social group cannot be understood unless it is seen in light of the continuing efforts of the individuals who create these conditions.

The marking of events—for example, the publications of books—attains significance only if they become "knots" (*Knotenpunkte*, Hegel, 1969, p. 435), that is, if there is an interpenetration between the progression of the individual and that of the social group. Only if these books are read, interpreted, and criticized do they become significant events in the course of the author's life. Conversely, they might also become significant happenings in the history of the social group if they influence a large number of impressed readers and if they generate new challenges and products. These knots are officially signified by historical dates, but these markings lie neither in the cultural-historical nor in the individual-psychological dimension alone but at the event points in which both intersect. Man creates history as much as he is created by history. The particular moments of effective interpenetration are recorded in the history of the society and in the curriculum of the individual.

4.4. Implications

In the development of their careers, scientists tentatively explicate their paradigmatic orientations through their teaching and their affiliation with a few like-minded colleagues and students. As they advance, they establish their own "scientific commune" and disseminate their ideas through reports and papers until these ideas become crystallized in research routines and textbooks. The small team with which they affiliate represents the nucleus of their activities and success. But the more they advance, the more they find recognition from other groups, which, though geographically remote, become attached to one another by shared knowledge and technology. Eventually, the larger group might represent a distinct paradigmatic orientation, which through its achievements creates a new constructive interpretation of a scientific theme.

As one cohort succeeds, it forces others into opposition and rejects earlier ones as insufficient. The resulting conditions create conflicts for coexisting orientations and crises for the preceding ones. These discrepancies between groups generate temporal markings and affect the individual members of competing groups. For a limited period, members of an older paradigm find some like-minded colleagues to lean on, but these persons might more successfully adjust to the changing conditions by moving more effectively forward in their individual scientific careers or by retreating from the laboratory into organizing and administrative duties. These events generate temporal markings that are experienced as crises. Crises can be resolved by structural transformations of the individuals' lives in concordance with the social progression. Ideally, one should outline a number of different structural progressions and implement them through the proper selection of assignments and allocation of resources. Such designs can be prepared only by careful observations of the temporal order of the scientific-cultural system wherein they take place. Therefore, the study of cultural-history progression has to be linked with the analysis of individual development. We can depict a developmental event-sequence only if we also describe the historical progression within which it takes place.

5. Dialectics, Development, and History

The preceding examples have given evidence for a concept of development that emphasizes event sequences whose temporal markings are generated by the interactions within or between single organisms, individuals, or social groups. The examples have also directed our attention to the biographical and retrospective study of life histories and historical changes. But the explorations of organisms, individuals, or social groups in their isolation do not provide comprehensive interpretations of their developments. Only the study of their interactive changes, in which points of conflict or conflict resolution mark the temporal order of the past, leads to such an understanding.

5.1. A Schema of Dialectical Changes

The proposed analysis has to be much further extended. In particular, I am proposing four dimensions of simultaneous changes: inner-biological, individual-psychological, cultural-sociological, and outer-physical (see Riegel, 1976a,b). If we consider the interaction of any one dimension with any other, we obtain the 4 × 4 matrix shown in Table 3.

The tentative labels attached to the 16 cells of Table 3 denote interactions with either a negative (upper word) or a positive (lower word) outcome. The cells along the main diagonal of the matrix represent conflicts and conflict resolutions within any particular dimension. They require some special elaborations. For example, in the cell denoting interactions within the event sequences along the inner-biological dimension, we might list at the molar level the cooperation or conflict between two organisms, for example, an attacking and an attacked animal, sexual partners, or mother and child. At the molecular

Table 3. Crises with Negative (Upper Lines) and Positive Outcomes (Lower Lines) Generated by Asynchronies along Four Planes of Developmental Progressions

	Inner-Biological	Individual-Psychological	Cultural-Sociological	Outer-Physical
Inner-biological	Infection Fertilization	Illness Maturation	Epidemic Cultivation	Deterioration Vitalization
Individual-psychological	Disorder Control	Discordance Concordance	Dissidence Organization	Destruction Creation
Cultural-sociological	Distortion Adaptation	Exploitation Acculturation	Conflict Cooperation	Devastation Conservation
Outer-physical	Annihilation Nutrition	Catastrophe Welfare	Disaster Enrichment	Chaos Harmony

level, we might think of the coordination (or lack of it) between different body organs, cell clusters, single cells, or cell body and cell nucleus.

Similar arguments have to be made, when one is exploring the interactions within any of the other three dimensions. For example, individual-psychological interactions include those between two people: husband and wife, parent and child, two friends, employer and employee, etc. At more specific levels, these interactions might involve subsystems within a single individual, for example, between the sensory and motor system, thought and speech, anxiety and motivation, or whatever categories psychologists are ready to propose. At still more specific levels, we need to consider the interaction between specific acts, behaviors, ideas, and expressions.

The constitutive groupings in the cultural-sociological dimension include the family, clan, tribe, nation, or civilization; the peer group, guild, union, party, or religious order; the neighborhood, community, or city; the age and sex group; generation; and many others. Equally wide is the distribution of outer-physical groupings. Here, we need to consider large-scale external interventions, such as earthquakes, floods, and droughts, as well as chemical and physical processes within organisms.

Undoubtedly, the matrix of comparisons is exceedingly wide, but there are no compelling reasons needed to consider all of them at the same time and for the same purpose. For any concrete analysis, selective restrictions are necessary. But whatever the choice of restriction, the event sequences to be explored are always generated by the interactions of at least two and, ultimately, of all dimensions of activities. The consideration of separated, single events is an abstraction that prevents any meaningful interpretations at the biological, psychological, sociological, and even physical levels.

In summary, I draw the following comparisons. In the inner-biological dimension, temporal markings appear as developmental shifts from activity to passivity, excitement to satisfaction, from growth to decay. All of them are determined in their interaction with psychological, sociological, and physical factors. In the individual-psychological dimension, temporal markings are experienced as resolutions and conflicts, decisions and doubts, movements and rests. In the cultural-sociological dimensions, temporal markings are objectified by incentives and restrictions, conquests and defeats, expansions and repressions. In the outer-physical dimension, temporal markings appear as night and day, summer and winter, floods and droughts. If we compare all of these dimensions with one another, the individual-psychological makes us aware of temporal markings; the cultural-sociological records events so that they can be transmitted to other people and to other generations. Because of their exceptional status, I devote further attention here to the exploration of the psychological and sociological dimensions and their experiential-individual and normative-cultural bases.

5.2. Cultural-Sociological Event Sequences

To the historians, reliable information is available on a selective basis, consisting only of what has passed through the hands of many generations of intervening participants. In trying to explore the transformation of historical information, I asked various groups of American students to write down in 10 minutes the names of historical figures influential in military, political, or governmental affairs (Riegel, 1973b). When I plotted the number of historical names listed against historical time, the results in Figure 6 were obtained. Since a large number of political figures were mentioned who entered history less than a few months prior to my study, a strong recency effect became apparent. The earliest accumulation of names occurred for the time of the American Revolution, most notably because of the frequent listing of George Washington. The absence of a strong primacy effect is understandable because history lacks a clear initiation or zero-point. These (American) students compensated for this lack, however, by considering the appearance of George Washington as the beginning of history, the "birth of a nation," that is, by relying on an intracultural-sociological conflict, the conflict between the British and the American colonists. Finally, Figure 6 shows the occurrence of sharp spikes, the first representing the time of the American Revolution, the others coinciding with outbreaks of wars and other major catastrophes.

Before I derived any further conclusions, I compared the records produced by students with different degrees of education and, presumably, historical knowledge, that is, freshmen, seniors, and graduates. I expected the more advanced students to show the spiking effect less strongly and to fill the gaps between the spikes more evenly with the names of historical figures not engaged in warfare and uprisings. However, this expectation was not confirmed. Although the spikes

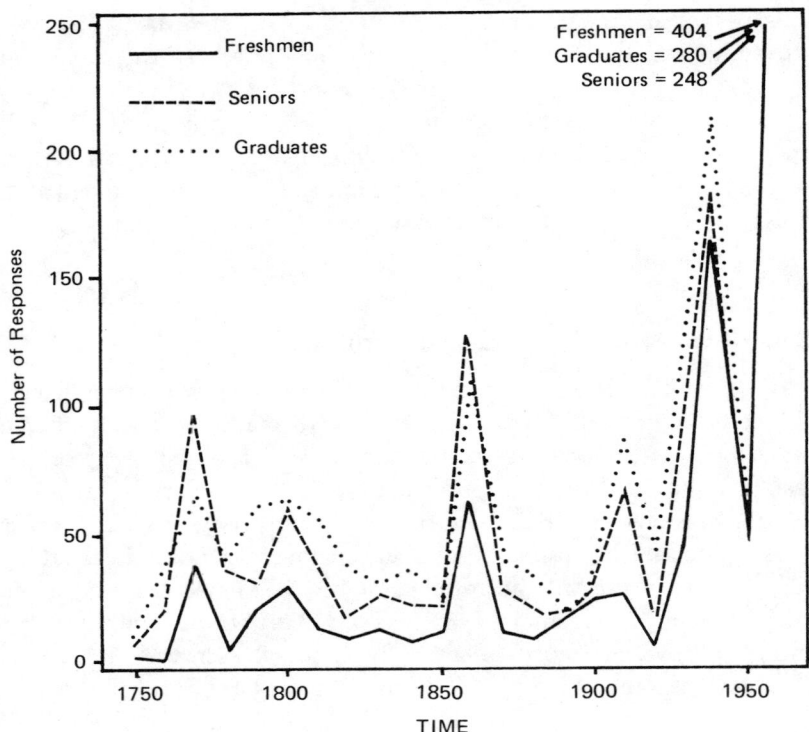

Figure 6. Number of persons influential in political, military, and governmental affairs named by three groups of 30 students each during 10-minute periods.

of the graduate students' records were less marked, and although one of the most formative periods in American history, the period at the beginning of the 19th century, was more evenly filled with names of historical figures, the differences between the student groups were not strong enough to provide convincing support for my expectation.

Next, I asked two other groups of students to list important persons in the areas of music, literature, and painting, hoping that I would obtain a similar, spiked distribution of names over historical time as observed for the political leaders, shifted, however, by one-half phase. In other words, I expected the peaks of the first task to coincide with the valleys of the second task. However, this expectation was not clearly confirmed either. When inspecting the distributions of the names of musicians and painters, I observed distinct peaks for historical periods intermediate between wars, that is, 1870, 1880, 1920, thus confirming the distributional-shift interpretation. For writers, however, large accumulations of names occurred at 1810, 1860, and 1940, that is, at periods of military instability.

After my first two attempts provided some suggestions but no definite conclusions regarding either selective biases in the recollection of historical names as a function of educational levels or the distributional shift as a function of military–political versus artistic–scientific dominations, I finally analyzed the most likely source of these biases, namely, the professional writing of political history. The results in Figure 7 were obtained from an advanced high-school book, *A History of the United States* by Alden and Magenis (1962), and show the number of lines given to each of the decades after 1750 in a summary of historical events, as well as the number of pages greater than two on which the names of historical figures appeared in the index of the book. The

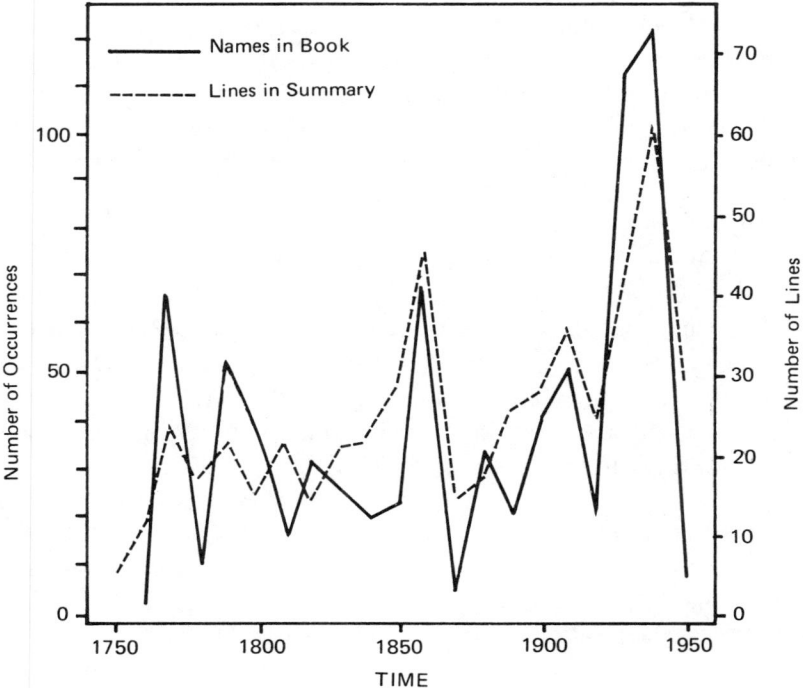

Figure 7.   Number of lines in a summary and number of names appearing in a book on American history.

results show more clearly than the listing of the names by the students that the writers of this book clearly emphasized military interventions and wars at the expense of advances in arts, sciences, education, and welfare. Very marked spikes are observed for the time of the American Revolution, the Civil War, and World Wars I and II.

5.3. The Dialectics of History

For a comprehensive interpretation of these results, let us consider the famous example of Caesar crossing the Rubicon during his march on Rome in the year 49 B.C. This event must have been quite accurately reported. After all, if a few days earlier Caesar was found to the north of the river and a few days later to the south, he could not have avoided the crossing. But regardless of how accurately these "facts" were recorded, their description, by itself, is insufficient to give them the status of a historical event. Only the interpretation of these steps by a historical perceiver as leading to civil war and the downfall of the earlier form of Roman government transforms these "facts" into historical events. Perhaps Caesar himself was aware of the potential interpretation; perhaps later admirers projected their own interpretation into the insignificant need to cross a small river in order to reach the destination, while Caesar himself remained unaware of its potential implications; perhaps the crossing was glorified in much the same way that Washington's crossing the Delaware was glorified in the famous painting by Emanuel Leutze and Eastman Johnson.

Our inability to learn "how it really was in history" (Ranke, 1885) should disturb us as little as our failure, according to Kant, to recognize "the thing as such." As shown in Figure 8, "history as it really was" is hidden behind a series of interpretive filters generated through the selective preservation of information by archivists, the insufficient scrutiny of scholars, the driving brevity of teachers, and the unchallengeable apathy of students. But even if we were able to look behind all of these filters, we would not find what we were hoping for because the events by and in themselves, in their numerousness and in their details—such as all those involved in the crossing of the Rubicon—are without historical meaning.

It is misleading, however, to denote these interpretations as selective filters. A filter selects "essential" issues from "unessential" details but presupposes that something is being filtered and recognized in its general features. Such an approach searches for "the historical thing as

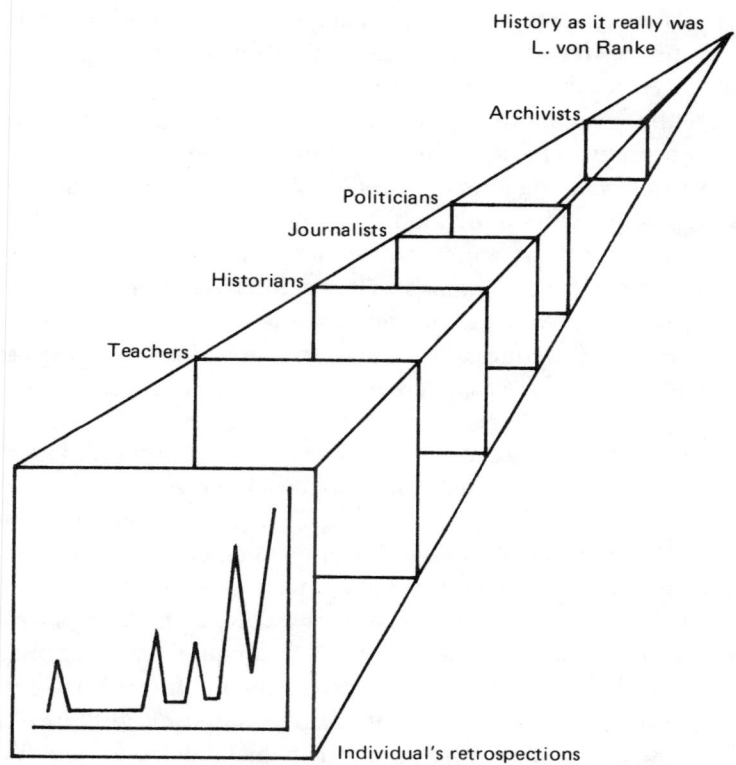

Figure 8.   Representation of historical schemata of interpretations.

such," the "facts" behind historical interpretations and claims to be "objective." Although few of us would deny that there are outer physical events, all that a historian can find behind a historical filter is another filter and another filter. If we do not wish to get lost in this regression *ad infinitum,* we must think of these selective filters as a sudarium.

A sudarium is a holy relic, for example the holy veil of St. Veronica, which when laid upon the face of the suffering Christ is said to have received his image forever. Every interpretation derived from a look at or through the sudarium imposes the image of Christ upon the events under concern. Unlike a history derived from the notion of historical filters, the concept of the sudarium proposed a "constructive" interpretation of history. It is predominantly concerned with the interpretation of sudaria and their relationships to one another. The systematic study of their cultural-sociological transformations and of the invariant properties sustaining these transformations represents the most advanced study of history. ◄

With the recognition of the constructive viewpoint, we apprehend both the normative character and the future dimensionality of history. As a person's development is directed by and based upon wishes, expectations, and hope at the individual-psychological level, so are historical changes determined by standards, values, and goals at the cultural-sociological level. If we recognize that our interpretations of history are dominated by apocalyptic views, emphasizing warfares and catastrophes at the expense of welfare, arts, and sciences, we have also gained insights and access to alternative conceptualizations. By exploring these options through exchanges, discussions, lectures, and writings, we generate and implement a new sudarium, a new interpretation of history, a new conception of man and his development. As emphatically claimed by Lynd (1968), the recognition of new interpretations and the awareness of former fallacies should lead us to an enactment of history.

Without any notable exception, psychologists have failed to apprehend the forward surge of knowledge. Like the "objective" historians, developmental psychologists, for example, have continued to screen the individual's behavior through their "filters" in order to describe "development as it really is." Endless and mostly futile efforts have been directed toward refining their methodologies and toward increasing the abstractness of their theoretical constructs. Developmental psychologists have failed to understand that the changes in experienced and lived development ought to be the most important topics of their analysis.

As much as the acquisition of advanced historical awareness begins to change history itself, so will an awareness of one's own development change the course of this development. Rather than apprehending our societal origin and our cultural history by the products generated, the viewpoint promoted here appreciates history by the activities that force it into new directions. The development of the individual, likewise, should no longer be apprehended by the products left behind, such as achievements and test scores but by the critical awareness of past experiences, which remain with the individuals and direct them toward their future. What we desire is neither a history of past failures— that is, catastrophes—nor a developmental psychology of petrified performances—that is, test records—but a science of the interactive development of the individual and society based upon lived experiences and directive actions.

5.4. Implications

In this section, I introduced some findings from the study of event sequences, presented some methodological suggestions for their analysis, and elaborated some theoretical issues in a dialectical analysis of time. The methodologies consisted in historical explorations applied to the study of the individual in the form of biographical inquiries or systematized in controlled recall-experiments. Both of these approaches need to be supported by sociological and anthropological investigations, and needless to say, both approaches require much further refinement and systematization.

History is generated by human organisms in their natural world and thus has its foundation at the inner biological and outer physical levels. By transforming these "lower" energies, man generates culture and history, which, in turn, transform man. Out of biological necessity, the human being invents, for example, supportive tools, communicative language, and social order. These products, in turn, transform the biological activities of human beings.

6. Space, Time, and Logic

The notion of universal, external dimensions—within which we believe we exist and which, in our search for knowledge, we believe we must recognize—has dominated our concept of time as much as it has dominated our concept of space. The notion of external time and space has undoubtedly attained extraordinary historical significance, but it burdens our investigations of behavioral and social processes rather than assisting in our understanding of them. In the concluding part of my chapter, I briefly review stages in the history of our concept of space, and by comparing these interpretations, I outline a dialectical concept of time.

6.1. Categorical, Relational, and Absolute Space

In the excellent treatise on the history of the concepts of space by Jammer (1954) and in several publications by Piaget (Piaget, 1970a,b;

Piaget & Inhelder, 1956), five stages in the history of the concept of space can be distinguished that correspond in general to the periods of cognitive development proposed by Piaget, that is, sensorimotor, pre-operational, concrete operational, and formal operational. The fifth stage of synthesis represents the dialectical period (Riegel, 1973a).

The first stage represents the Greek conception of space, sometimes called the Aristotelian space concept. Here, the primary focus was on substances or objects. Space was regarded as filling the gaps between these substances; space was regarded negatively as a void. In contrast, our modern conception gives priority and independence to space; objects are considered to be located in space and can be defined and measured within particular spatial systems. The distinction between these two orientations has been pointedly explored by Cassirer (1910). Similarly, in Piaget's view, the young child's concept of space is topological. Children see or grasp objects here and there without being able to derive a generalized notion of space. Insofar as they stretch their arms toward objects or move from object to object, their space concept has become relational.

The second stage leads to the relational concept of space as origi-nally developed in Arabic philosophy, most notably by Al-Ghazālī (Algazel, 1927), and introduced into modern philosophy and science by Leibniz and Huygens in one of their major confrontations with Locke and Newton. According to Jammer (1954), "Leibniz rejected Newton's theory of an absolute space on the ground that space is nothing but a network of relations among coexisting things. In his correspondence with Clarke, Leibniz likens space to a system of genealogical lines, a 'tree of genealogy' or pedigree, in which a place is assigned to every person" (p. 48).

Also in the child, the relational space concept supercedes and elaborates the concept of substance, object, or categorical space of the young infant. As children look at objects and move their sight from one object to the next, a relational space concept emerges. But in spite of a successful subject–object differentiation (Riegel, 1975c), the child remains an integral part of the spatial system. The relations connecting the observing child with object A and B and his or her movement from A to B eventually yield three connected points that form the basis of a primitive plane. Any further movements and fixations by the child define additional planes and more complex relational structures. The relational space may also be conceived of as consisting of concentric spheres defined by the movements of various objects around the observing or manipulating child. This onion-layer model of space eventually becomes the Cartesian system of coordinates. Most notably, in both versions of the relational space, the observer remains a central part and an anchor point of the system.

Two operations, which Piaget has called *decentration* and *seriation,* support the development of the space concept at the third stage and lead to a conceptual extension of the relations. Thus, the vector connecting, for example, the observer and the object is extended both forward and backward across a transitive series of objects. The geometry of the resulting space, projective geometry, is historically much younger than the geometry appropriate for the space concept of the next stage, Euclidean geometry. In the history of painting and design, however, the techniques for representing space projectively have been known for many centuries, at least since the early Renaissance.

As Piaget has shown, concrete operational children (7–11 years) are able to perform decentration by moving conceptually from their own position to those of other observers, for example, at either side or opposite. As demonstrated by the "three-mountains problem," the older child might judge correctly whether the mountains cover each other when seen from the positions of different observers. Thus, the self-centered concept of young children becomes flexible and is no longer bound to their particular locations.

By the conception of projection lines as defined by the observer and an object or by two different objects, a constructive representation of space emerges that is not bound to the conditions here and now but attains a generality like the representation of the three-dimensional space on the two-dimensional canvas of the late Renaissance painters. In paintings, the two (and occasionally three) projection (vanishing) points are externally given; commonly, they lie on a line representing the horizon. The projection lines of the psychological observer, in comparison, diverge like rays from his or her pupils. Thus, a decisive step in the abstraction from the human observer is achieved.

In spite of all these advances, the projective space remains concrete, that is, observer-centered. But the observer can now take various positions, even imaginary ones. The important shift toward the concept of classical Euclidean or absolute Newtonian space during the fourth stage of development consists in nothing else than providing an ideal position for an ideal observer, that is, moving him or her infinitely far away from the observed objects, onto God's lap. Thereby, the crucial eighth axiom of Euclidean geometry is met, namely, that parallel lines intersect in infinity, an axiom whose sense is hard to apprehend unless viewed with this particular spatial metaphor in mind. If the observer is moved infinitely far away, the projective lines that either converged upon the vanishing points on the canvas or diverged from the eyes of the observer have become parallel. Thus, a formal operational description of space has been achieved that relies on classical mathematics, especially on analytical geometry. Such a space can never be perceived (except by the observer on God's lap); one can only think about it. But the purely imaginary character also means that it is ideal and remains immutable from observations.

Our thinking has been dominated to such an extent by the concept of an absolute, formal operational space that it caused considerable difficulties to modern natural scientists and philosophers to overcome this orientation. Philosophers came to regard space as an externally given container in which all objects and events were located and came to regard time as a mysterious fourth dimension. Rather than focusing on the distribution and order of substances as the Greeks had done, philosophers reduced these substances to microscopic particles (atoms or points) whose movements through space and time were explored in the classical laws of the natural sciences. Philosophers and scientists either came to regard space, time, and substance as external properties of nature (Locke's primary qualities) or as properties of the mind, which impose law and order upon nature (Kant's *a priori* forms). As elaborated by Cassirer (1910), the concept of substance became less important than the concepts of time and space, especially that of space.

Modern sciences and philosophy have not remained fixed upon the concepts of absolute space and time but, since the later parts of the 19th century, have either tried to reduce time to space (as in thermodynamics) or to fuse time, space, and/or substances (as in relativity and quantum theory). Thus far, the behavioral and social sciences have failed to keep up with these important developments. In this chapter, I have tried to elaborate some modern options, especially as they relate to the concept of time, which I summarize and extend in the following section.

6.2. Relational, Absolute, and Dialectical Time

In Greek philosophy, space was negatively defined as the domain that was not filled with substances (objects); space was regarded as a void. In comparable manner, time could be regarded as the stretch that is not filled with events; thus, time would be negatively defined by events. However, events imply action and thus reflect a higher degree of organization than substances in space. A concept of time based on a series of events has been called *relational*.

Relational time arises from serial interactions between events. Like a simple monophonic melody or a passage in a narrative, it is produced by the contrast between preceding and following sounds and sound groupings. Temporal structures of individual experiences and cultural documentations likewise arise from variations of event sequences. Since these event sequences usually involve the movement of objects (or substances) in space, relational time compounds the three basic concepts of classical natural sciences and philosophy: substance, space, and time. Activity is its most basic property.

In comparison to relational time, which is intrinsically defined either in the experience of the observer or by the overt order of the events, absolute time is extrinsically defined and "regarded as constitutive of nature . . . it regulates the rates of processes . . . [and] is physically prior to events" (Fraser, 1967, p. 835). It represents the most abstract form of conceptualization. In music, for example, it is superimposed by the meter and the bar lines on the score sheets or by means of the clicks of the metronome during rehearsal.

Finally, dialectical time is comparable to polyphonic compositions and thus incorporates a diversity of monophonic or relational sequences. It also incorporates absolute time because polyphonic music too relies upon an extrinsic standard—especially for the synchronization of different instruments or voices—that represents one particular monophonic structure that has been elevated to serve this function. Dialectical time is both intrinsically and extrinsically determined and thus indicates that any experience of time involves the interaction of at least two event-sequences, for example, the phenomena observed and the measurement taken.

Dialectical time emphasizes concrete experiences and events. As these lead to conflicts and resolutions, questions and answers, disharmonies and harmonies, temporal markings are produced by the synchronization of these event sequences. These markings, knots, or points of coincidence represent transitions in the sequence of changes. Harmonies and disharmonies represent momentary states in a flux of changes. They resemble structures that lack temporal extension and are similar to semantic fields in distinction from syntactic orders. These harmonies and disharmonies merge into the temporal organization of melodies. The contrastive comparison between simultaneous spatial conditions and developmental temporal changes elucidates the basic properties of dialectical time.

At the beginning of this chapter, I discussed the interactions between event sequences at the inner biological level. But for the most part, greater attention has been given to the temporal structures of individual experience and cultural representation. These interactions lead to the identification of temporal markings when based upon sequences of actions and thoughts within individuals or to the recording of historical dates when these temporal structures represent interactions between different cultural groups.

Temporal structures emerging through the interactions of cultural groups create conditions that appear as "objective" and "normative" to the individuals. Subsequently, these conditions exert a stronger influence upon the individual than the individual can exert upon the culture. In music, for example, the general style of a historical period—the melodic forms, rhythm, and meter—might impose upon the individual an order that appears to be objective and firm. Creative advances represent attempts to break with these traditions. In science and philosophy, these normative orders may reflect a particular theoretical orientation prevailing in the culture at a given historical time period, for example, the absolute time concept of Newton, Locke, and Kant.

Throughout their histories, societies have introduced a variety of different time concepts. Most cultures have been dominated by categorical and cyclic concepts of time (Nakamura, 1966; Needham, 1966; Russell, 1966). If the relationality of time was emphasized, it commonly found expression in genealogical orders and hierarchies. Later in history, calendar time was introduced, and only a few centuries ago was introduced the formal concept of "absolute time," which represents the most abstract conception and the hardest to overcome. With the emphasis upon event sequences and their interactions, I have supported a multidimensional though concrete concept of time, that is, a concept based upon the common experience of individuals and the social relations between them.

The interactive basis and the emphasis upon concrete experience reveals the dialectical character of this time concept. As event sequences are interwoven with one another, temporal markings are produced by their synchronization. As harmonies and disharmonies in polyphonic musical compositions merge into the temporal organization of melodies, these temporal markings represent momentary states in a flux of changes. The contrastive comparison between simultaneous spatial conditions and developmental temporal changes, finally, carries our discussion into logic and mathematics.

6.3. Formal and Dialectical Logic

Traditional logic, as well as traditional science and philosophy, has been exclusively concerned with nontemporal, static conditions. The conception of a nonchanging state of being originated with Eleatic philosophy and was thought to reflect the universal order of the cosmos. Astronomy (as well as astrology), logic, and mathematics are our heritage from the adevelopmental and ahistorical thinking of Eleatic philosophers.

In his studies of music, Pythagoras began to explore temporal relations. Being restricted to monophonic music and the simultaneous (that is, nontemporal) relations between sounds, his studies were limited, however, and aimed at the "detection" of a universal, atemporal order in musical expressions. Extensive studies of temporal organization emerged only when complex temporal structures of polyphonic orchestration began to dominate musical compositions.

Nevertheless, our concept of time has remained rather limited. Our sciences, founded upon formal logic, deal exclusively with spatial structures. Even when Galileo investigated the laws of gravity and Newton the law of motion, the procedure consisted in measuring the spatial conditions at some time slices and then to infer the temporal relations. Only when dialectical logic began to be developed at the beginning of the 19th century did a model of thinking emerge that was fitted for the explication of change, development, and history. Polyphonic musical compositions, in designing complex temporal organizations, prepared the way for such a conception.

Dialectical logic is more than a counterpart to formal logic. Just as the movements in music are built upon momentary slices in the sequence of changes, so does dialectical thinking presuppose formal logical structures. Dialectical logic recognizes that it cannot exist without formal logic. This recognition provides a more general basis for dialectical logic than is available to formal logic. Formal logic fails to recognize such mutuality and is bound to consider itself immutable. Dialectical logic represents an open system of thinking that can always be extended to incorporate more restricted systems. Formal logic aims at a single universal analysis. As a consequence, it cannot apprehend itself. In particular, it cannot apprehend itself in the developmental and historical process.

The development of dialectical logic, especially by Hegel, has not been commonly accepted in the natural sciences (at least not in the classical natural sciences), and most surprisingly, it has received less attention in the behavioral and social sciences. Since—as it has been convincingly demonstrated by Kosok (1976)—dialectical logic can be cast into a systematic language that makes it applicable to the sciences, and since dialectical logic is the mode of thinking that alone can deal appropriately with change, development, and history, such a disregard or neglect is regrettable indeed. Nevertheless, both the rapidly growing appreciation of dialectical logic in the natural sciences and the similar, if belated, recognition of its significance among behavioral and social scientists[3] indicate that a decisive change is in the making.

References

Alden, J. R., & Magenis, A. *A history of the United States.* New York: American Book Co., 1962.

Algazel. *Tahafot Al-Falasifat.* M. Bouygens (Ed.). Beirut: S.J., 1927.

Baltes, P. B. Longitudinal and cross-sectional sequences in the study of age and generation effects. *Human Development,* 1968, *11,* 145–171.

Birren, J. E. Principles of research on aging. *In* J. E. Birren (Ed.), *Handbook of aging and the individual.* Chicago: University of Chicago Press, 1959. Pp. 3–42.

Boring, E. G. *The history of experimental psychology.* (2nd ed.) New York: Appleton-Century, 1957.

---

[3] The topic of a dialectical psychology has recently found considerable support among behavioral and social scientists. As it is in a state of constant change, a comprehensive list of publications cannot be provided. The following list includes some of the recent books and articles. Books: Buss, 1977; Datan & Reese, 1977; Harris, 1977; Riegel, 1975b, 1976c; Riegel & Rosenwald, 1975; Rychlak, 1976. Articles: Buck-Morss, 1975; Buss, 1975; Freedle, 1975; Fourcher, 1975; Harris, 1975; Hefner, Rebecca, & Oleshansky, 1975; Lawler, 1975; Meacham, 1972, 1977; Meacham & Riegel, 1977; Mitroff & Betz, 1972; Riegel, 1973a, 1976a,b; Sameroff, 1975; Van den Daele, 1975; Wozniak, 1975a,b.

Buck-Morss, S. Socio-economic bias in Piaget's theory and its implications for the cross-cultural controversy. *Human Development,* 1975, *18,* 35–49.

Buss, A. R. The emerging field of the sociology of psychological knowledge. *American Psychologist,* 1975, *30,* 988–1002.

Buss, A. R. (Ed.). *The social context of psychological theory: Toward a sociology of psychology knowledge.* New York: Irvington, 1977 (in press).

Cassirer, E. *Substanzbegriff und Funktionsbegriff.* Berlin: B. Cassirer, 1910. (*Substance and function and Einstein's theory of relativity.* Chicago: Open Court Publishing Co., 1923.)

Clive, J. (Ed.). *Thomas Carlyle: History of Frederick the Great.* Chicago: University of Chicago Press, 1969.

Datan, N., & Reese, H. W. (Eds.). *Life-span developmental psychology: Dialectical perspectives of experimental research.* New York: Academic Press, 1977.

Dobbs, H. A. C. Time and extrasensory perception. *Proceedings of the Society for Psychical Research,* 1965, *54,* 249–361.

Fieguth, R. *Struktur des literarischen Wandels: Structur der Einzelwerke.* Unpublished manuscript, University of Konstanz, Germany, 1973.

Fourcher, L. A. Psychological pathology and social reciprocity. *Human Development,* 1975, *19,* 405–429.

Fraser, J. T. The interdisciplinary study of time. *Annals of the New York Academy of Sciences,* 1967, *138,* 822–847.

Freedle, R. Dialogue and inquiry systems: The development of social logic. *Human Development,* 1975, *18,* 97–118.

Freedle, R. Human development, the new logical systems, and general systems theory: Preliminaries to developing a psycho-social linguistics. *In* G. Steiner (Ed.), *Piaget's developmental and cognitive psychology within an extended context. Vol. 7. The psychology of the 20th century.* Zurich: Kindler, 1977.

Gunter, R. *Sentences in dialogue.* Columbia, S.C.: Hornbeam Press, 1974.

Harris, A. E. Social dialectics and language: Mother and child construct the discourse. *Human Development,* 1975, *18,* 80–96.

Harris, A. E. *Dialectics: A paradigm for social sciences.* 1977 (in preparation).

Hefner, R., Rebecca, M., & Oleshansky, B. Development of sex role transcendence. *Human Development,* 1975, *18,* 143–158.

Hegel, G. W. F. *Wissenschaft der Logik.* Frankfurt/M.: Suhrkamp, 1969. (*Science of logic.* London: Allen & Unwin, 1929.)

Hoagland, H. The physiological control of judgments of duration: Evidence for a chemical clock. *Journal of General Psychology,* 1933, *9,* 267–287.

Hoagland, H. The pacemaker of human brain waves in normals and in general paretics. *American Journal of Physiology,* 1936, *116,* 604–615.

Hölder, O. Die Axiome der Quantität und die Lehre vom Mass. *Berichte der Sächsischen Gesellschaft der Wissenschaften, Leipzig, mathematische-physikalische Klasse,* 1901, *53,* 1–64.

Husserl, E. *The phenomenology of internal time consciousness.* The Hague: Nijhoff, 1964.

Jammer, M. *The concepts of space.* New York: Harper, 1954.

Kosok, M. The systematization of dialectical logic for the study of development and change. *Human Development,* 1976, *19,* 325–350.

Kvale, S. The temporality of memory. *Journal of Phenomenological Research,* 1974, *5,* 7–31.

Labov, W., & Waletzky, J. Narrative analysis: Oral versions of personal experience. *In* J. P. Helm (Ed.), *Essays on verbal and visual arts.* Seattle: University of Washington Press, 1967. Pp. 12–44.

Lawler, J. Dialectic philosophy and developmental psychology: Hegel and Piaget on contradiction. *Human Development,* 1975, *18,* 1–17.

Lewis, M., & Freedle, R. Mother–infant dyad: The cradle of meaning. *Research Bulletin 72-22.* Princeton, N.J.: Educational Testing Service, 1972.

Lynd, S. Historical past and existential present. *In* T. Roszak (Ed.), *The dissenting academy.* New York: Pantheon Books, 1968. Pp. 101–109.

Mao Tse-tung. *Four essays in philosophy.* Peking: Foreign Language Press, 1968.

Meacham, J. A. The development of memory abilities in the individual and society. *Human Development,* 1972, *15,* 205–228.

Meacham, J. A. Soviet investigations of memory development. *In* R. V. Kail & J. W. Hagen (Eds.), *Perspectives on the development of memory and cognition.* Hillsdale, N.J.: Erlbaum, 1977.

Meacham, J. A., & Riegel, K. F. The final period of cognitive development: Dialectic operations. *In* G. Steiner (Ed.), *Piaget's developmental and cognitive psychology within an extended context. Vol. 7. The psychology of the 20th century.* Zurich: Kindler, 1977.

Meerloo, J. *Along the fourth dimension.* New York: John Day, 1970.

Merleau-Ponty, M. *The phenomenology of perception.* London: Routledge & Kegan Paul, 1962.

Mitroff, I. I., & Betz, F. Dialectic decision theory: A metatheory of decision making. *Management Science,* 1972, *19,* 11–24.

Nakamura, H. Time in Indian and Japanese thought. *In* J. T. Fraser (Ed.), *The voices of time.* New York: Braziller, 1966. Pp. 77–91.

Needham, J. Time and knowledge in China and the West. *In* J. T. Fraser (Ed.), *The voices of time.* New York: Braziller, 1966. Pp. 92–135.

Piaget, J. *The language and thought of the child.* New York: Harcourt & Brace, 1926.

Piaget, J. *The child's conception of movement and speed.* New York: Ballantine (paperback), 1970a.

Piaget, J. *The child's conception of time.* New York: Ballantine (paperback), 1970b.

Piaget, J., & Inhelder, B. *The child's conception of space.* London: Routledge & Kegan Paul, 1956.

Popper, K. Letter to the editors: Irreversible processes in physical theory. *Nature,* 1958, *181,* 402–403.

Ranke, L. von *Geschichte der romanischen und germanischen Völker von 1494 bis 1515* (3. Aufl.). Leipzig: Duncker & Humblot, 1885.

Reichenbach, H. *Philosophie der Raum-Zeit-Lehre.* Berlin: Gruyter, 1928. (*The philosophy of space and time.* New York: Dover, 1958.)

Reichenbach, M., & Mathern, A. The place of time and aging in the natural sciences and scientific philosophy. *In* J. E. Birren (Ed.), *Handbook of aging and the individual*. Chicago: Chicago University Press, 1959. Pp. 43–80.

Riegel, K. F. Age and cultural differences as determinants of word associations: Suggestions for their analysis. *Psychological Reports,* 1965, *16,* 75–78.

Riegel, K. F. Development of Language: Suggestions for a verbal fallout model. *Human Development,* 1966, *9,* 97–120.

Riegel, K. F. Some theoretical considerations of bilingual development. *Psychological Bulletin,* 1968, *70,* 647–670.

Riegel, K. F. Time and change in the development of the individual and society. *In* H. Reese (Ed.), *Advances in child development and behavior.* Vol. 7. New York: Academic Press, 1972. Pp. 81–113.

Riegel, K. F. Dialectic operations: The final period of cognitive development. *Human Development,* 1973a, *16,* 346–370.

Riegel, K. F. The recall of historical events. *Behavioral Science,* 1973b, *18,* 354–363.

Riegel, K. F. Contrastive and recursive relations. *Research Memorandum RM-74-23.* Princeton, N.J.: Educational Testing Service, 1974.

Riegel, K. F. Adult life crises: Toward a dialectic theory of development. *In* N. Datan & L. H. Ginsberg (Eds.), *Life-span developmental psychology: Normative life crises.* New York: Academic Press, 1975a. Pp. 97–124.

Riegel, K. F. (Ed.). *The development of dialectical operations.* Basel: Karger, 1975b.

Riegel, K. F. Subject–object alienation in psychological experimentation and testing. *Human Development,* 1975c, *18,* 181–193.

Riegel, K. F. The dialectics of human development. *American Psychologist,* 1976a, *31,* 689–700.

Riegel, K. F. From traits and equilibrium toward developmental dialectics. *In* W. J. Arnold & J. K. Cole (Eds.), *1974-75 Nebraska Symposium on motivation.* Lincoln: University of Nebraska Press, 1976b. Pp. 349–407.

Riegel, K. F. *The psychology of development and history.* New York: Plenum Press, 1976c.

Riegel, K. F. The dialectics of time. *In* N. Datan & H. W. Reese (Eds.), *Life-span developmental psychology: Dialectical perspectives on experimental research.* New York: Academic Press, 1977. Pp. 3–45.

Riegel, K. F., & Rosenwald, G. C. (Eds.). *Structure and transformation: Developmental and historical aspects.* New York: Wiley, 1975.

Russell, J. L. Time in Christian thought. *In* J. T. Fraser (Ed.), *The voices of time.* New York: Braziller, 1966. Pp. 59–76.

Rychlak, J. F. (Ed.). *Dialectic: Humanistic rationale for behavior and development.* Basel: Karger, 1976.

Sameroff, A. Transactional models in early social relations. *Human Development,* 1975, *18,* 65–79.

Schaie, K. W. A general model for the study of developmental problems. *Psychological Bulletin,* 1965, *64,* 92–107.

Schmid, H. *Anthropologische Konstanten und literarische Struktur.* Unpublished manuscript, University of Bochum, Germany, 1973.

Stevens, S. S. Mathematics, measurement, and psychophysics. *In* S. S. Stevens (Ed.), *Handbook of experimental psychology.* New York: Wiley, 1951. Pp. 1–49.

Testa, A. *The dialogical structure of language.* Bologna: Cappelli, 1970.

Van den Daele, L. Ego development in dialectic perspective. *Human Development,* 1975, *18,* 129–142.

Volosinov, V. N. *Marxism and the philosophy of language.* New York: Seminar Press, 1973.

Wozniak, R. H. A dialectic paradigm for psychological research: Implications drawn from the history of psychology in the Soviet Union. *Human Development,* 1975a, *18,* 18–34.

Wozniak, R. H. Dialecticism and structuralism: The philosophical foundation of Soviet psychology and Piagetian cognitive developmental theory. *In* K. F. Riegel & G. C. Rosenwald (Eds.), *Structure and transformations: Developmental and historical aspects.* New York: Wiley, 1975b. Pp. 25–47.

Wundt, W. *Völkerpsychologie, Bd. 1 & 2, Die Sprache.* Leipzig: Engelmann, 1911–1912.

Editors' Introduction

In everyday life, social and personal events are rarely seen in isolation; they acquire their full meaning from their contexts. We ask ourselves, "What's the story?" in order to assess the meaning of events in our lives. We tend to look at an event as possessing a beginning, a middle, and an end. However, since there is nothing in physical reality that is called a "beginning" or an "end," we have to define these terms through the social frameworks whereby people attribute beginnings and endings to events. In this chapter, Stuart Albert and William Jones address themselves to the techniques people use to define the temporal limits of social interactions.

In several previous articles, Albert has investigated the construction of the social endings and the concluding rituals in conversations, meetings, and wars. In this chapter, Albert and Jones sharpen their focus and examine the structuring of endings in a common situation: the telling of a child's bedtime story.

Like Kastenbaum, Riegel, and Schlesinger, Albert and Jones demonstrate how the interplay of a number of ongoing events and structures leads to a common outcome; in this case, the act of getting a child to sleep. By using a method called *template composite analysis,* Albert and Jones demonstrate that the different, yet coordinated, components of the bedtime story lead to an inevitable ending. By considering cases in which the structural rules of the story are violated, they show that poorly structured stories do not end conclusively and that the children who hear these stories will probably not be led to sleep.

Albert and Jones's template analysis methods provide promising tools for the analysis of social time and the coordination of events.

The Temporal Transition from Being Together to Being Alone: The Significance and Structure of Children's Bedtime Stories

Stuart Albert and William Jones

A focus on the temporal dimension of social life immediately reveals two fundamental categories—being together and being alone—and the fact that much of our life is spent in making the temporal transition between the two: between being alone and coming into contact with others, or being with others and leaving their presence to be either alone or with someone else. This paper focuses on the temporal transition from being together to being alone, in short, on the problems of ending social encounters. In other papers, we have developed the importance of the study of endings in general (Albert, 1973, 1975; Albert & Kessler, 1976a,b). Let us note here only that endings between individuals are frequent, that at times they are very difficult to achieve without intense anxiety and emotional turmoil, and finally, that they are central to the study of change that is conceptualized as a temporal partition between the end of one encounter, relationship, or process and the beginning of another.

In this paper, we are interested in the construction of a particular

Stuart Albert and William Jones • Department of Psychology, University of Pennsylvania, Philadelphia, Pennsylvania.

ending, one that is culturally universal, biologically necessary, and that occurs with daily frequency. We are interested in the ending that occurs between a parent and a child when the child is put to sleep.

The social-psychological features of putting a child to sleep are of particular interest because it is during this time that the child is taught how endings ought to occur, what properties a good ending ought to have, and how an ending is to be accomplished. It is also an important occasion for defining what it means to be with others and what it means to be alone. Our interest is to provide a social-psychological account of this transition. Elsewhere (Albert & Kessler, 1976a,b), we have described a general set of psychological processes that we hypothesize as occupying the temporal boundary between being together and being alone. In this paper, we apply some of those processes to the analysis of a particular ending. But the central objective is to add to the study of human endings an analysis of a particular symbolic invention, one that adults employ to aid them in constructing a satisfactory ending with a child. This symbolic invention or object is a children's bedtime story. We now proceed to conceptualize a bedtime story as an ending device and then develop a method for analyzing the internal structure of a story so that we can see how this structure serves important ending functions.

1. A Bedtime Story as an Ending Device

For our purposes, a bedtime story consists of a portable and re-usable set of psychological processes that are prepared in attractive artistic and literary form and that can be purchased inexpensively in the marketplace by any member of the culture. Our task here is to provide an account of what these processes are and how they function as ending devices.

Our analysis is of stories that already exist in book form, although the method of template-composite analysis to be developed can also be applied to stories that are spontaneously generated. Typically, an analysis of bedtime stories might be expected to proceed directly to considerations of structure and content, plot and characters. There is an even more basic level of analysis, however, that has been left totally unexplored. This level of analysis derives from the fact that a written story and the storybook that contains it are symbolic and physical objects. We argue here that certain properties of a bedtime story as a symbolic and physical object make it useful as an ending device.

1.1. Properties of a Bedtime Story as a Symbolic and Physical Object

1. Usually, although one can listen to something even while one is actively engaged in doing something else, it is easier to listen when one is still rather than physically or cognitively active, and a story is precisely the sort of object that produces a state of quiet as a precondition for its receipt. Moreover, since a story is pleasurable, the child is willing to be still and listen. By comparison, "wake up" or "good morning" stories do not exist. They would be a contradiction in terms, for the requirement of awakening is to energize the organism and not to ensure that it is put into a resting state. Indeed, society has rules of politeness that require that the audience convey the impression that they are listening by being quiet and by looking in the appropriate place. If the eyes of the audience wander, we commonly believe that their minds are wandering. The listening individual, therefore, has the task of conveying the fact of his listening. If he does not, he may be subject to censure. Communicating the proper degree of attention and hence respect—that is, presenting evidence of having assumed the appropriate posture of the listener—is made easier for the child by the physical existence of a book as something that can be looked at. Not only does looking at the pictures or the text of a book allow the child a sense of participation, of being able to follow along even though he cannot read, but the desire to observe the book also confines him to a relatively small physical space and this confinement moves him one step closer to a state of quiescence. Thus, a story is a gift, the receipt of which is governed by social rules requiring that the recipient diminish his activity as a condition for receiving the gift.

2. The story has an end that is physically coordinated to the end of the book, unless of course the book contains many stories. But even here, there is direct physical evidence of the place where one story stops and another begins in the size of the type, the layout of the page, etc., even without a consideration of the formulaic nature of beginnings and endings within the text itself. The fact that the story and the storybook are physical objects that have clear spatial boundaries present evidence to the

child that the end of the story is not imposed solely by the adult for the adult's convenience but has an independent reality as well. While it is perhaps possible to argue that the story ought to have ended earlier or later, the existence of the book as a finite and bounded physical object is an authoritative statement about where it in fact does and does not begin and end.

There is often a blank page at the end of a book or a blank inside cover in the book. The existence of these blank, unwritten-on spaces is clear evidence that the story is over. (We are here, of course, talking about a book with only one story in it.) The story did not end for want of paper on which to print it, and simply because space was available as a container on which to place a string of additional words did not mean that this space had to be filled. The lesson to the child by analogy is that just because time exists there is no necessary reason to fill it with waking activities. In the broadest sense, our claim is that an upper temporal boundary to the social encounter between parent and child is facilitated and in part constructed by the adult's use of a finite and bounded spatial object. In any case, the words "The End" are sometimes appended to make clear that the issue is not in doubt.

Putting the child to sleep is an exercise in the rights and the treacheries of authority. The adult has the power to leave the child, to have him exiled to his proper place, and the child has the will to protest with all the resources at his command. Evidence that the adult is also subject to outside authority—in this case, the authority of an unseen author who has brought the story to a close—provides a useful model for the child, who is also being asked to accept an externally imposed limit. In effect, the adult states that he is willing to accept an external definition of an ending, and so should the child. (The child, of course, may ask the adult to read another story, but that, of course, is a different problem.)

3. A story is a source of reward and punishment. It is pleasurable. It therefore has the status of a gift that, if it is freely given, conveys to the recipient a sense of being cherished and valued; if given contingent on going to sleep, it implies a principle of exchange. The adult states or implies that he will read the child a story and that in return, the child will go to sleep. Moreover, like anything valuable, a story can be withheld as punishment.

4. When an adult reads a child a story, they both enter and share a common fictional world and are united in experiencing its pains and pleasures. The bedtime story, therefore, is a physical and symbolic

object around which a relationship of pleasurable sharing can develop. It is a tangible locus for the expression of a social bond: it can be physically held by two individuals at the same time. All this fits well with the observation that many endings are characterized by increased attraction, cohesiveness, and intimacy between individuals (Goffman, 1971; Albert, 1973; Albert & Kessler, 1976a).

5. The story is also an object around which complementary social relationships are developed. It is the child's book, but it is the adult who reads it. The child gives his attention as an audience, the adult his best efforts as a performer. The child must decide whether to solicit the story as a bribe for going to sleep. And once the bribe is offered, he must decide whether to allow himself to be corrupted.

6. The story itself contains both fictional and real elements. Therefore, it is the perfect object to occupy the transition state between the waking world and the world of dreams. To become engrossed in the story is to let go of the reality of the setting in which the storytelling is taking place. This letting go is a certain kind of ending in the restricted sense of a shift of attention away from the adult who is reading the story to the world of the characters in the book. Furthermore, the story represents the delights of the world of the imagination, and imagination is precisely the kind of human faculty that can be exercised by a solitary individual. Thus, the adult brings to the child an example of the kind of cognitive activity that can exist without the presence of others.

7. The existence of an imaginary world also establishes the point that it is possible for interesting interpersonal events to occur even if the people are not actually physically present in the child's bedroom. A story is something that the child can think about when the adult leaves. It is therefore a psychological gift given by the adult as a substitute object of attention, attention that might otherwise be directed to the loss of the adult. In this sense, the story serves an important continuity function (see Albert & Kessler, 1976a,b). While the physical encounter between parent and child will end, the fruits of that encounter will continue as a source of pleasure to the child.

8. A story has movable parts (namely, pages) in fixed order (the book is bound), so that it is possible to know what part uniquely follows the next part, and moreover, it is possible to be assured that no part is missing. Considered as a physical object, the book subdivides the narrative into discrete sections, and thereby provides some support for the idea that a discontinuity of symbolic experience is quite natural.

9. The fact that the story is partitioned into pages has a more important implication, however, and that is that the story is only progressively revealed to the child. We would argue that the progressive unfolding of a story serves in part as a demonstration that there is a future even if the future is not seen or known. (A bedtime story presented to the child on one large page might be psychologically less effective in this regard.) In addition, the child learns not only that something is coming next, but with increasing repetition of the story, he learns to predict exactly what is coming next and will often correct the adult if the adult deviates from the text.

We might posit as one of the design requirements of a satisfactory ending that it demonstrate over and over again that if one is at time 1, there, in fact, will be a time 2. Of course, that demonstration would have little effect if the child didn't care whether there was a time 2 or not. But a successful story captures the child's attention and he does care what comes next. In the symbolic device of a story, therefore, we have a psychological situation in which the child wants to know what is coming next, and the nature of the story is such that we never disappoint him.

The unfolding nature of the story takes place without any effort on the part of the child. Having read page 15, he need do nothing in order to receive the delights of page 16. The importance of establishing faith that a future will take place even though the child does nothing to make it happen is vitally important if the child is to learn that sleep is an interruption and not a final ending. Once asleep, he will not see or hear what happens in the world he has left, a prospect that is no doubt disappointing, but while he may fear that he may miss something, the greater fear is that the world itself will stop when he goes to sleep, that there will be nothing coming next, that the conversation, once interrupted, will never continue. The repetition of a known linear sequence involving fictional characters that matter to the child is part of the logical apparatus necessary to construct the idea of continuity and hence the concept of interruption.

Repetition may seem foolish to us, but that is because we are already assured of the permanence of the social world and of the fact that events unfold in time even when we are asleep. But to a child, it is just those properties of the world that are in doubt and that are being learned, and since the issue at stake—namely, one's own survival—is so great, it is perfectly reasonable to expect children to prefer to hear the same story over and over again. (Of course, there are, no doubt, other reasons for repetition as well.) While a belief in object constancy may be acquired at a very early age, the belief that a social relationship may endure even when the participants are not present may be acquired much later.

The above analysis suggests that flashbacks and flashforwards, even if they are understood by the child, should be largely absent from the literature of bedtime stories, as they disrupt the smooth flow from past to future that we have argued is essential to a belief in an unseen future.

10. Bedtime stories almost always have people in them or, if not people, animals with the characteristics of people. This does not appear to be a startling observation as most stories have people in them. Yet, consider an alternative, a story about various kinds of bookcases, for example. Our claim is not simply that bookcase stories are a boring genre but that bookcases would be singularly inappropriate as the symbolic contents of a temporal boundary that constitutes an interpersonal ending. Such a boundary calls for the presence of people and of interpersonal relationships. The empirical claim is that bookcase stories would not work as well in putting the child to sleep.

11. We would be remiss if we did not mention as the final property of the bedtime story as a symbolic and physical object one of the most obvious considerations. The story, if it is long and not overly exciting, provides the adult with sufficient resources to outlast the child, to bore, lull, or otherwise simply exhaust him until he falls asleep. Interestingly enough, even when the child falls asleep in the middle of the story, an adult sometimes continues to the end of the story if it is near, for even if the child has achieved a satisfactory ending, the adult has not.

We turn now to an examination of the internal structure of children's stories. There is, of course, a vast literature illustrating a variety of perspectives (see Bettelheim, 1976, for the most recent popular treatment of children's fairy tales). Some authors pursue a psychoanalytic orientation. Others derive notions from linguistics and seek a grammar of children's stories that will enable them to understand how we remember certain stories and not others (see Mandler *et al.*, 1976). Others examine stories within the context of sociolinguistics or ethnomethodology (see Sacks, 1972). Our aim is quite specific and limited. We are not interested in a comprehensive understanding of children's stories, to which there are many approaches. Rather, our goal is to develop an analysis of a bedtime story that is useful in understanding the story as an ending device; that is, we seek a method of analysis that will allow us to understand the story's role in the construction of a particular temporal boundary.

Our orientation is that there are certain problems in constructing an ending to a social encounter and that certain properties of a bedtime story are useful in solving these problems. This approach was implicit in our previous enumeration of the properties of a story as a symbolic and physical object. We now consider how the internal structure of a bedtime story can contribute to the construction of an ending. We term this method of analyzing the internal structure of a story *template composite analysis*.

2. Template Composite Analysis

Template composite analysis is best developed in the context of a concrete example. For that purpose, we relate the child's bedtime story entitled *Ten Bears in My Bed: A Goodnight Countdown* written by Stan Mack (1974). The story opens with a small boy coming into his bedroom only to find that there are 10 bears in his bed. There is a look of surprise and mild displeasure on his face. He begins to order the bears out of his bed, one by one. The bears in the bed show some dismay at being ordered to leave, but each departs through the window of the child's bedroom, happily playing with one of the child's toys. The story is an adaptation of a familiar round, as follows:

> *There were Ten Bears in My Bed*
> *And the Little one Said*
> *Roll over, Roll over*
>
> *So they all rolled over,*
> *And one flew out* [of the room].
>
> *There were nine in his bed* [etc.] .[1]

Finally, all of the bears have departed. The little boy then takes a teddy bear that was in the corner of the room into bed with him. His mother peers into the room, says goodnight, and puts out the light. On the final page of the book, the little boy is seen dreaming of all the bears happily

[1] Copyright 1974 by Pantheon. Reprinted by permission.

playing with his toys, which have magically reappeared in his room.

Let us analyze this story in terms of a set of logical structures or templates. The first template in Figure 1 is a descending counting structure starting with 10 and proceeding to zero and then to plus 1. The bed contains 10 bears, and they leave, one by one, until the little boy gets in. The properties of a descending counting algorithm as a logical format for the story is that once the story has this form, an ending is ensured by the properties of the number system. It is not simply that we know when the series will end with a sense of completeness, namely at zero, but that we also know the rules for getting there. If we were inserted blindfolded into any point of the story, we would immediately be oriented. We would know what came before, and what is likely to come next, and what direction we have to travel to get to the end, and how far away the end is. Therefore, not only is an ending assured by a descending counting structure, but a considerable degree of coherence is given to our movement in that direction. This template, which solved the ending problem in advance as soon as it was accepted, is coordinated with a set of other templates.

The second template we label *rest–tired,* although we might have chosen other labels for this process of fatigue. While the bears do not put up any great protest in having to leave the bed (perhaps bribed by the opportunity to leave with the child's toys), it is still clear that some work is involved in getting the bears to leave. Imagine if there were 40 bears in the bed, and the little boy still had to extricate them one at a time. The effort might be more than he could bear, and we might well find him asleep on the floor with a dozen or so bears still left in bed.

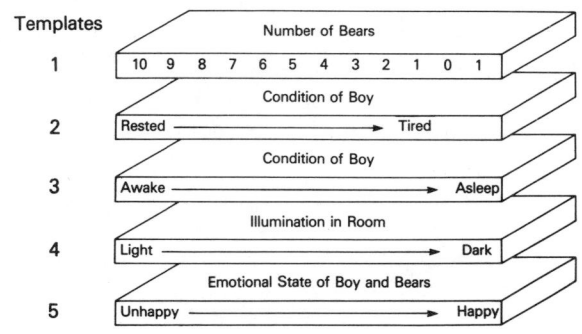

Figure 1.   A template-composite analysis of *Ten Bears in My Bed.*

The lesson is clear. In order to go to sleep, it is helpful to be tired and to have earned the rest as a result of meaningful achievement.

The third template is given by the terms *awake–asleep* and refers to the little boy.

The fourth template refers to a property of the physical environment and is given by the terms *light–dark*. The light in the room remains on until it is turned off by the little boy's mother.

The story has an emotional structure as well. The fifth template is referred to by the terms *unhappy–happy*. It applies both to the little boy and to the bears collectively. The little boy is startled and somewhat displeased to find all the bears in his bed and therefore orders them out. The bears, in turn, are unhappy at being displaced from the bed. In the boy's dream at the end of the story, however, all the bears are seen happily playing with the little boy's toys.

We now see the story as a lamination or superimposition of at least five distinct logical structures or templates. Let us examine the way the different templates are coordinated with each other.

2.1. Template Synchronization

One of the best ways to find the coordination of the templates is to examine some of the ways this coordination could be disrupted. Suppose, for example, that the little boy fell asleep on the floor when there were still four bears left in the bed. Or suppose that the last bear had left the bed but that for some reason, the little boy delayed getting into bed and going to sleep. This possibility is illustrated in Figure 2. In either case, we would say that the first and second templates are not synchronized and that the stories that result from template synchronization are less satisfactory. This is an empirical issue on at least two counts. First, do others including children share this judgment, and second, would asynchronous stories of this kind be less successful in a pragmatic sense—would they be less successful in putting the child to sleep?

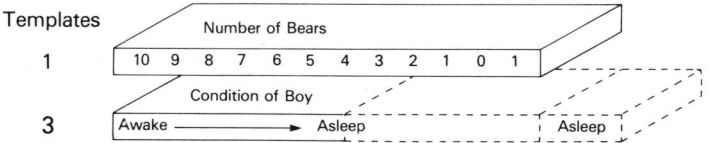

Figure 2.   An example of template asynchronization.

Consider another asynchrony. Suppose the child became noticeably happy when there were four bears left in the bed or always remained unhappy even when he went to sleep, only becoming happy after he had been dreaming for some time. Again, we would argue that the ending of the story would lose some of its force.

The degree of template asynchrony can be defined in terms of the number of templates that are not synchronized with each other and the degree to which they are out of phase, the latter effect made possible because the first template has 12 possible points of coordination to which the other two-point templates can become attached.

A story can be disrupted not only by template asynchronization but by what we call *template inversion*. Consider the templates referred to by the terms *awake–asleep* and *light–dark*. Each template is an ordered pair of attributes, so that the second attribute is usually characteristic of or associated with an ending. It is possible (although it does not always make sense) to write a story with a template inversion in it in which the term usually associated with ending is placed first rather than last. For example, one could imagine a story in which the room was dark at the beginning of the story but illuminated at the end. Or one could imagine a story in which everyone started out happy but ended up unhappy.

In addition to template inversion, it is possible to add or delete templates. We refer to the *thickness* of a story as the number of super-imposed templates it contains. Thus, the total conceptual machinery supplied by template composite analysis is template addition, deletion, synchronization, and inversion. We should state that although we present here no discovery procedure for finding templates, we subsequently present a series of tests for determining the independence of one template from another once each has been identified.

In order for the ending of the story to have the necessary speed and definiteness, there must be sufficient force for an ending to occur, in other words, sufficient template thickness, synchronization, and lack of inversions.

Finally, two additional processes are important in bringing about a sense of closure: summarization and the assertion of continuity. The operation of these processes is presented in some detail elsewhere (Albert & Kessler, 1976a). A summary is defined as the selective repetition of elements from the past history of the encounter represented in a condensed and symbolic form. In this story, the history of all the bears leaving is summarized in the little boy's dream, in which all of the departed bears are again reassembled. The summary is selective in the sense that only instances of happiness are included. The unhappiness at having to leave the little boy's bed is omitted. The summary process therefore contributes to a happy ending by being precisely the kind of information-processing device that scans the past history of the encounter and assembles selective instances of positive affect. For a full analysis of the summary process of ending, the reader is referred to Albert and Kessler (1976a).

There are several important statements of continuity. The first is the appearance of the boy's mother. The boy falls asleep with the assurance that an enduring relationship with his parents has been affirmed. Thus, to end an encounter at the end of the day, a belief is needed that there is an overarching relationship that is not severed by physical disengagement and change of conscious state. The expression of continuity cannot be excessive, however. Imagine if the whole family gathered at the child's bedside. The child might begin to wonder. An overly strong assertion of continuity may raise the very fears it seeks to quiet.

A second expression of continuity is found in the child's dream of all the bears playing with his toys. The child is instructed that if he is willing to part from the bears, he will be reunited with them, and not only that, but the reunion will be pleasant.

The final source of continuity is the symbolic substitition of a toy bear for the departed real bears. The little boy takes a symbolic bear to bed with him, tangible evidence that the bears have not really left.

The story of the 10 bears is, of course, no longer as simple as it appeared, constructed as it is from a complex set of synchronized templates together with at least two additional processes of ending, summarization, and assertion of continuity. Its surface simplicity has been complexly achieved.

3. Problems of Infinity: An Ascending Counting Algorithm

If our analysis is correct, reversing the descending counting algorithm and making it an ascending counting algorithm should create enormous problems of ending, for one would have an infinite series rather than a sequence converging to a known limit. If such a story were to achieve an end at all, it would have to use more powerful devices. In fact, we have found such a comparison story. Let us present the story and then its analysis, which will then allow us to specify more precisely some features of template composite analysis.

The story is the classic tale of the *Five Chinese Brothers* written by Claire Huchet Bishop and Kurt Wiese (1973). It opens with a picture of five Chinese brothers who look exactly alike but who have special skills and characteristics. The first Chinese brother can swallow the sea, the second has an iron neck, the third can stretch his legs indefinitely, the fourth cannot be burned, and the fifth can hold his breath as long as he likes. The first brother sets off to the ocean to go fishing, accompanied by a little boy from the village who has promised to obey the first brother explicitly as a condition of being allowed to go along. The first brother proceeds to swallow the sea and gather in the fish, which are now easily caught. The little boy who accompanied the first brother plays happily on the sea bed but, alas, does not return when the first brother calls him. The first brother, unable to hold the sea in his mouth any longer, is finally forced to let it return, thus drowning the little boy.

When the first brother returns to the village, he is accused of drowning the little boy, and he is tried, convicted, and sentenced to have his head cut off the next morning. However, the first brother implores the judge to let him return home allegedly to say goodbye to his mother. In reality, this is just a ploy to allow the second brother to be substituted for the first.

The result of this substitution is that the attempted execution fails, since the second brother has the iron neck. The judge then sentences the brother to death by drowning. The same ploy, however, allows the second brother to go home and the third brother to return in his place. Since the third brother's legs can stretch forever, he cannot be drowned. Similar attempts at execution by burning the brother and by smothering him in an oven of whipped cream also fail because the fourth brother, who is substituted, cannot be burned; neither can the fifth brother be smothered, since he can hold his breath indefinitely.

As each execution fails, the villagers become increasingly angry. While each attempted execution begins during the morning and is completed that day, during the last attempted execution, the villagers are said to stay outside the oven all day and all night, as the story relates, "just to make sure."

Finally, as the fifth brother emerges from the oven unharmed, the judge declares that since all attempts at an execution have failed, it must be that the brother is innocent. The villagers agree and the brother is allowed to return to his family, where it is said that *he and his family lived happily for many years.* Notice that they do not live happily forever after, but only for many years. The story ends with something less than complete resolution, and the difference between living happily forever after and living happily only for many years is part of what we want to explain.

3.1. A Template Composite Analysis

We present a schematic diagram of the templates that make up this story in Figure 3.

The first template is an ascending counting structure that lists the number of ways an execution can be carried out. The significant feature of this structure is that there are an infinite number of ways of killing someone. Notice that this template is merely an inversion of the first template in the analysis of *Ten Bears in My Bed.*

The second template refers to the number of Chinese brothers and is a five-point template. In Gestalt terminology, the five brothers form a family, or a unit, and a certain force toward closure is achieved by an enumeration of each element of the set.

The relationship of the first two templates establishes the central problem of how the story is to be ended; that is, the story must stop

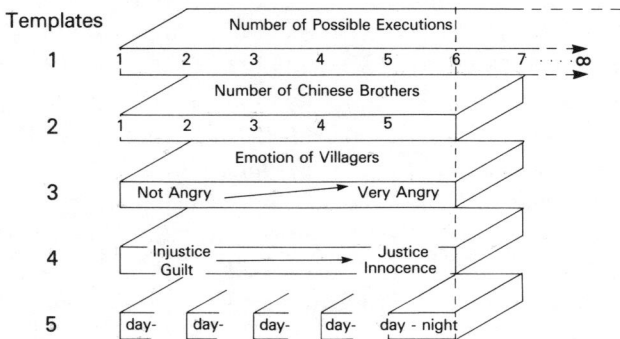

Figure 3.   A template-composite analysis of *The Five Chinese Brothers.*

when the last Chinese brother has been substituted, although there are still an infinite number of ways in which executions could be carried out. How is the problem to be solved? It is solved, we argue, by the addition and synchronization of other templates, each of which contributes some force to the ending. These templates are put in phase with the substitution of the last Chinese brother.

The third template is an emotional one. When the first execution fails, the story relates that the villagers are simply angry. When the second attempt fails, the villagers are described as very angry, and when the third attempt fails the villagers are described as getting more and more angry every minute. The implication of this template is that there is a limit to becoming angry without something's happening. We cannot say in advance how angry a person or a group of persons can become before the anger is simply unleashed in some way, but we believe that an increase in anger cannot go on forever, and it is precisely this notion of an implied limit that allows this template to contribute to a force toward closure.

The fourth template is one concerned with ethics and morality. The Chinese brother is declared guilty at the beginning of the story, an attribution of guilt that the reader knows is to some degree, at least, unwarranted. An injustice must be corrected. When it is, the story can end. Justice denied but then achieved provides a sequential structure to the story that contributes an immensely powerful sense of closure. Although we are not able to explore the issue in detail here, we feel that it is no accident that moral judgments tend to occur at the end of stories rather than during the middle, for example, and that the definitions of what is good, desirable, and ethical are talked about as the proper *ends* of life.

To claim that justice has been achieved presupposes that all pertinent information has been assembled and evaluated. An ethical stand is undertaken only with reluctance, if at all, unless there is some assurance that all relevant information is at hand. (This requirement sometimes implies the prior operation of the summary process, the function of which is to scan the history of the event or encounter for relevant information and to reassemble that information in a condensed form for all to examine. A trial, for example, assembles all relevant evidence prior to a judgment of guilt or innocence.) When an ethical statement is advanced, there may, in fact, be things in the story that are still vague, or incomplete, but whatever they are they do not bear on the critical issues at stake. Thus, the claim that justice has been achieved is the claim that the story is complete.

Whether in fact such a claim of justice will be believed and/or accepted when it is advanced depends on whether the audience feel they know all the things that are relevant, which implies that the audience or the readers have some implicit theory of what constitutes a complete description of behavior relative to some theory of ethics. How people decide that they know everything they need to know about a situation—how they detect omissions—is an empirical question that remains to be investigated. In this story, the claim of justice is advanced and probably accepted in part because all of the Chinese brothers have been accounted for.

The final template is a day–night cycle. Each execution begins on a new day. During the last attempted execution, the story relates that the villagers stay outside the oven all day and all night, thus completing the day–night cycle and thereby giving the story a sense of closure.

This template also implies a process of physical exhaustion, as the villagers have been up all night. We may expect them to be tired and want to rest, a state that is typically associated with an ending. There is at the same time a sense of cognitive exhaustion at being unable to think up any more ways to get rid of the Chinese brother. Thus, contained in the day–night cycle are two kinds of exhaustion, both physical and mental, the presence of which contributes to a sense of closure.

Each of the templates we have discussed adds its own force toward closure, and we suggest that these forces add together when the templates are superimposed on each other. Some templates contribute a stronger sense of closure than others, but our claim is that they all add something. However, this incremental force toward closure occurs only to the extent that the templates are synchronized with each other.

3.2. Template Synchronization

One of the best ways of considering the topic of template synchronization is to examine several instances of asynchronization. Suppose, for example, that the anger of the villagers reached a clear peak before the end of the story and then declined. Or suppose that it was discovered that the original Chinese brother was innocent at the time of the second or third execution. A reluctant eyewitness may have finally decided to come forward, for example. Clearly, the stories generated by these asynchronies would be less satisfactory. Their endings would be flawed.

Our claim is that despite the collective force toward closure that is marshaled by the synchronized superimposition of templates 2–5, the problem of infinity contained in the first template is never completely overcome, for it is always possible for the villagers to think of yet another way to execute the Chinese brother. For that reason, the brothers are said to live happily for many years, but perhaps not forever.

We should add that even if the reader claims that there is no real distinction between living happily forever after and living happily for many years or that the latter ending occurred in this story only as an accident perhaps due to the author's desire to add some variety to the ending, the claims of template composite analysis permit independent validation, for the assertion is that adding and subtracting templates will affect the sense of closure that is achieved by the story.

3.3. Living Happily Forever After

No discussion of the construction of temporal boundaries could be complete without a discussion of the phrase "and they all lived happily forever after." Let us unpack the significance of each of these terms for what such analysis can tell us about the desired properties of an ending.

The word *all* means that despite whatever separations have occurred in the course of the story—in this case, the separation of brother from brother—provided the characters have not died, they will be reunited with each other at the end. Happiness, therefore, is collective. Stories rarely end with the major characters separated or isolated from each other but nonetheless happy.

Consider another alternative to the emotion of happiness. Suppose a story ended, "And they lived with exquisite and intense joy for ever and ever." We would consider that ending somewhat implausible because we believe that different emotional states have different degrees of permanence. Some, such as joy, are believed to be short-lived. Of all the emotions, it is possible that happiness is the longest-lasting positive emotion. That is, we propose that the product of the desirability of happiness times its expected duration is greater than the rated desirability of some more positive emotion times its expected duration. Other states may be preferred, but we don't expect them to last, and others may last, but they are not preferred. This proposition is, of course, a testable one.

There is another feature of happiness beyond what one might call its expected "half-life" and that is its implication of privacy. When we are told that someone is happy, it is not only that we believe that his life is probably uninteresting (a belief that may or may not be fueled by jealousy) but that we do not wish to inquire too closely about it. It is intrusive and in bad taste to closely interrogate an individual who claims that he is happy. This is true not simply because his happiness depends on not being pestered no matter how pure the motives of the interrogator but because we fear that the state of happiness would not survive close examination. Happiness examined is happiness destroyed. Thus, the claim that happiness has been achieved is equivalent to a claim of privacy and an admonition to the reader that we have no special right to invade it; and if we cannot inquire further, the story must end at this point.

Finally, there is a temporal dimension to the phrase "and they lived happily forever after." That is the assertion that nothing will change in the future, the claim that the same homogeneous state endures forever. But what makes experience interesting is in part its variability. There is a vast literature that documents a preference for at least some moderate level of environmental and cognitive complexity.

4. Summary and Conclusions

Our goal in focusing on certain properties of a bedtime story and our purpose in developing the method of template composite analysis is not to contribute to literary criticism and to the analysis of folklore but to provide a social-psychological account of how a bedtime story may serve as an ending device. Although template composite analysis is set within the general analytic tradition developed by Goffman in his pioneering work on frame analysis (Goffman, 1974), it is much more limited in scope.

The contribution of template composite analysis is twofold. First, template composite analysis provides an account of what makes the ending of the story one that a child can imitate and/or experience as relevant to his own condition. We needed an analysis of successful story endings, but successful in a restricted sense: successful not in terms of theories of literary criticism, but successful in dealing with the kinds of problems the child himself is facing. One of the principal problems for the child is that the time he must go to bed is arbitrary and almost invariably seen as too early. What is needed is a way to analyze the ending of a bedtime story to see what contribution the story makes to the successful resolution of the child's problem. The concept of synchronization deals with the problem of the arbitrariness of the child's ending by describing how a stimulus is constructed that illustrates the experience of a nonarbitrary ending. In part, what we mean by a "natural ending" is one in which the ending of one thing occurs at the same time as the ending of another. If we leave a baseball park when the game is over, that leaving is considered natural; to leave earlier or later is not. Synchronization of action is not the only criterion of naturalness, but it is surely one criterion. In effect, just at the time a child may feel that his bedtime is arbitrarily imposed by authority out of phase with his own wishes, the authority comes forward with a present that contains an illustration of synchronous arrival. The adult and the child become absorbed in telling and listening to the story so that the achievement of synchrony—only one component of a satisfactory ending to be sure, but perhaps a necessary one—occurs during the time when the issues of arbitrariness are potentially troubling.

The second contribution of template composite analysis comes from the idea of *thickness,* defined as the number of templates that compose a story. The principal idea is that each template contributes a force toward closure and that by adding templates, provided they are properly synchronized, one adds power to the ending. There are several things we need to examine about this notion. The first is a limitation. We do not have a discovery procedure for identifying templates. We have provided an analysis of two cases, but we do not have a dictionary of templates and a set of rules for applying them to the analysis of the individual cases, although such a dictionary and set of rules may ultimately be achieved. Moreover, there are different kinds of templates, and we cannot know now exactly what the best classification is. For example, some templates deal with the enumeration of finite sets, others with the idea of biological exhaustion. Only the detailed study of those stories that are actually used at bedtime can provide us with clues about what kinds of templates there are and how best to classify them.

If we cannot specify how a story is to be described in terms of a set of templates, does not the notion of thickness lose its force? One analyst might find 10 templates, while another might find only 5. Fortunately, we are not in this position, for having proposed a set of templates, we can specify a series of test procedures to determine whether the templates are independent of one another.

In summary, the central psychological problem that is solved when a story has adequate thickness is whether there is a sufficient reason for the story to end. (Here we might add that thickness might be better defined as a weighted sum rather than as simply the number of templates, since some templates contribute a stronger force toward closure than others.) Having sufficient reason for a termination is also part of the problem of going to bed for the child. Template composite analysis presents a characterization of the stimulus that is made available to the child as a bedtime story in such a way that we can see how the stimulus provides a symbolic or imaginary experience of sufficient causation.

Since a determination of template independence is necessary if the notion of thickness is to have predictive utility, we outline three tests that are useful.

1. *The test of synchronization and asynchronization.* To determine whether two templates are independent it is necessary to show that they can be put in and out of phase with each other: when they are synchronized, one story results, and when they are not synchronized, another story results.

2. *The test of omission.* The second test of independence between two templates is whether one can be omitted from the story and the other one retained. Consider, for example, templates 1 and 2 in the story of the 10 bears. There would still be a story if there were no bears in the bed at all and the little boy simply went into his bedroom and went to sleep.

3. *The test of addition.* A final test of the independence of two templates is to insert each into another story and show that when each is inserted a different story is thereby generated.

There may, of course, be other tests that can be developed for template independence, but it is clear that when applied to a provisional set of templates, these procedures of synchronization, omission, and addition greatly clarify their relationship to each other. Moreover, what has become evident is that template-composite analysis is not simply a method for analyzing the endings of existing stories but is itself a fertile means of generating new stories. We can envision a cycle of stories based on all possible states of template asynchronization.

We should mention that while we have developed the method of template-composite analysis within the context of an existing children's story, its potential range of application is much broader, for it can be applied to ongoing social interaction as well. Here one might examine how individuals actively construct certain templates so as to bring about an ending. In a field observation of children's bedtimes, Krakow (1976) reported one incident in which parents had their children running around the house in order to tire them out so that they would go to sleep. This is a conscious attempt to generate the template of exertion and rest that we saw in the story of the *Ten Bears* and that was implied in the day–night cycle of the *Five Chinese Brothers*.

4.1. Concluding Comments

The social ending we have examined is coordinated with our biological clock and is a part of all human society. While going to sleep is a biological necessity, how it is done and how we disengage from the social world are acquired skills. From the anthropological point of view, going to sleep (leaving the presence of others to enter another state of consciousness) may be one of the most important rites of passage for human beings, one that perhaps even serves as a prototype for other rites of passage (see Van Gennep, 1960) that occur less frequently during the life cycle of the individual. The psychological importance of going to bed as a rite of passage is due not to those infrequent occasions in which it is publicly celebrated but to its almost unnoticed embeddedness in the routines of daily life.

While some bedtime stories may appear playful and light and evoke a sense of pleasure, the psychological issues with which they deal are deep and potentially troubling: the problems of legitimate authority; of fear of separation, and of being left, abandoned, exiled, coerced, and compelled; of knowing that one will not die but merely sleep; of knowing that one will continue to be cared for and protected; and so on. The innocence of the bedtime story belies the seriousness of the problem for which it is in part an illustration of a solution.

References

Albert, S. Dynamics and paradoxes of the ending process. Paper presented at the annual
    conference of the Center for Research in Cognition and Affect, City University of
    New York, New York, 1973.
Albert, S. The study of endings as a temporal place. Paper presented at the American
    Psychological Association symposium: Historical and Temporal Issues in Social
    Psychology, 1975.
Albert, S., & Kessler, S. Fundamental processes in the construction of the ending of
    social encounters: Some empirical evidence. Unpublished manuscript, University of
    Pennsylvania, 1976a.
Albert, S., & Kessler, S. Processes for ending social encounters: The conceptual ar-
    cheology of a temporal place. *Journal for the Theory of Social Behavior,* October
    1976b, *6,* 147–170.
Bettelheim, B. *The uses of enchantment: The meaning and importance of fairy tales.*
    New York: Alfred A. Knopf, 1976.
Bishop, C. H., & Wiese, K. *The five Chinese brothers.* New York: Scholastic Book
    Services, 1973.
Goffman, E. *Relations in public: Microstudies of the public order.* New York: Basic
    Books, 1971.
Goffman, E. *Frame analysis: An essay on the organization of experience.* New York:
    Harper and Row, 1974.
Gumperz, J. J. & Hymes, D. (Eds.). *Directions in sociolinguistics: The ethnography of
    communication.* New York: Holt, Rinehart & Winston, Inc., 1972.
Krakow, M. Unpublished field notes on bedtime rituals, 1976.
Mack, S. *Ten bears in my bed: A goodnight countdown.* New York: Pantheon Books,
    1974.
Mandler, J. M., Johnson, N. S., & DeForest, M. A structural analysis of stories and
    their recall: From "once upon a time" to "happily ever after." La Jolla: University
    of California at San Diego, Center for Human Information Processing, March,
    1976.
Sacks, H. On the analyzability of stories by children in sociolinguistics. *In* J. Gumperz
    & D. Hymes (Eds.), *Directions in sociolinguistics: The ethnography of communi-
    cation.* New York: Holt, Rinehart, & Winston, Inc., 1972.
Van Gennep, A. *The rites of passage.* Chicago: The University of Chicago Press, 1960.

4

Editors' Introduction

In this chapter, Gilbert Voyat poses the question of whether time is a basic, immediate, and intuitively "given" aspect of reality or whether, instead, time is derived from more primitive concepts. We believe that we "know" the equation by which *time = distance/velocity*. However, velocity itself is often defined in terms of time. Voyat leads us out of this potential tautology by illustrating how children acquire the notion of time without necessarily having to use the concept of velocity *per se*.

Voyat shows how young children employ quite different referents for judgments of time than do adults. For example, young children equate "more time" with "more speed," while other children and adults realize that "more speed" implies less time. In experiments in which space, speed, and travel time were independently varied, Voyat shows that until the child is about 9 years old, the concepts of space, speed, and time are not coordinated in any comprehensive operational equation. However, by 9 years of age, most children's notions of time are constructed from velocity and distance judgments.

Since time is a cognitive construction and never is directly experienced by the child or the adult, Voyat contends that many temporal distortions can be understood by the fact that people may be confusing "time" with one of its underlying referents. From these insights, Voyat draws some interesting implications for research in children and adults.

Perception and Concept of Time: A Developmental Perspective

Gilbert Voyat

1. Introduction[1]

The content of this paper reflects a twofold preoccupation. First, the epistemological nature of time is a complex problem. As a concept, it is simultaneously defined in itself and, more often interpreted through its relationship with space, movement, and speed (Piaget, 1971).

In common terms, one understands time vaguely as something moving toward the future or as something in which events point toward that direction. Many contradictions are embodied in this understanding. These contradictions have led philosophers to postulate various epistemological doctrines, many of which are concerned with the concept of time. Some of them have had as a specific aim the elimination of some of the problems implied by this "common-sense" point of view. What is this future toward which events point? In short, how can time embody changes and permanencies simultaneously?

In this context, the first famous but unresolved controversy arose in ancient Greece between Paramides, who maintained that changes were irrational illusions, and Heraclitus, who affirmed that there was no permanence and that changes characterized everything without excep-

[1] In memoriam A. Cole, A. Gilbert, and H. Perl, three irreplaceable friends who died last year.

Gilbert Voyat • Department of Psychology, City College, City University of New York, New York.

tion. Centuries later, another controversy arose between the disciples of Newton and of Leibniz. According to the first, time was independent of, and prior to, events. In his own words, "absolute time and mathematical time of itself and from its own nature flows equally without regard to anything eternal" (Newton, 1739, p. 12). On another hand, according to Leibniz, time is not independent of events because time is formed by events and the relations among them. Therefore, it constitutes a universal order of succession.

This controversy gave rise to the doctrine of space–time, in which both space and time are considered as two systems of relations distinct from a perceptual point of view but inseparably bound together (Sivad-jian, 1938).

As one can readily deduce from this rapid overview, time, as an epistemological concept, entails more than covert psychological components. Although from an external and perceptual point of view one can, in principle, distinguish space and time, from a formal and structural point of view space and time entertain relationships (not the least one being the possibility of expressing one by the other; see Piaget, 1970) in which their distinction is not an easy task.

Many facets of this controversy have led thinkers to believe that the concept of time could not be fully accounted for unless one distinguished between perceptual or subjective time and objective or conceptual time. The first one is confined to the perceptually shifting "now" of the present, whereas the second includes all periods of time in which the events we call past, present, and future can be mutually related.

This distinction leads to the second aspect of time, time as a psychological concept. As P. Fraisse (1957) noted when analyzing its historical development,

> the common feature of all the authors dealing with time, until the middle century derives from the fact that they attempt to interpret our idea about time from an analysis of our state of consciousness. . . . This reflexive pathway simultaneously psychological in its analysis and philosophical in its aim, is still actual and can be considered as a permanent attribute of the human mind (pp. 7–8).

From this point of view, the problem of distinguishing space and time arises.

P. Janet, who introduced a new perspective on this problem, distin-guished space as "the most common and simplest form of this eternal multiplicity, to which we react by changes of position" and time which, "as another aspect of this multiplicity, involves a set of feelings, a set of phenomena sustained by the regulations of our actions" (1928, p. 26). According to him, time is a reaction from our consciousness toward the experience of change: time constitutes the feelings of change, and consequently time is, in its content, a result of consciousness. Thus, the most primitive (primitive in a developmental sense) behavior relative to time refers to the behavior of effort, from which derives a sense of duration. Yet this feeling is not a pure action but a regulation of actions that results from the necessity to adapt ourselves to irreversible changes. As P. Fraisse (1957) noted, "the psychological problem is not to know what time is, not to understand the nature of our idea about time, not to capture its genesis as an intuition or reconstruction in our mind, but to understand how one reacts to a situation embodied in changes" (p. 10). The problem here, then, is to understand the place that actions (including internalized actions) occupy when a subject provides a judgment of time.

2. Problems Related to the Development of Time

At the level of the subject, the place assigned to the actions implies (after the appearance of the symbolic function) an activity of mental reconstruction. This activity is particularly crucial in the case of time, primarily because of its irreversible nature. In effect, this reconstruction leads to three crucial problems: first, the problem of the relationship between time, space and speed; second, the question of the relationship between time and spatial coordinations; and third, the problem of the nature of time itself—is it a construction resulting from a coordination of space, of accomplished work and speed, or is it a simple and direct intuition?

From a developmental point of view, to solve these problems is of great importance. Its resolution could lead to some possible solution to the apparently vicious circle that is, in principle, posed from a formal standpoint by the interplay of space, time, and speed.

To clarify this issue, let us consider, for instance, the paradox that some philosophers have come to when defining them. In order to define time, one needs a watch. What is a watch? It is an apparatus that generates a spatial displacement at a constant speed. What is speed? It is a distance during a unit of time. The various attempts at defining time all tend to return to a simple circle: time equals space over speed $(t = d/s)$ or speed equals space over time $(s = d/t)$. The difficulties of the philosopher in extracting time from other concepts conceived of as simpler might very well be related to the fact that in the child, time as an abstract concept does not evolve through an elementary process from a set of time behaviors or an array of practical sensorimotor concepts previously established.

In this respect, Piaget (1970) has shown that the development of the concept of time depends upon a progressive decentration, that it implies an active participation of the subject, that, initially, there is a lack of differentiation between spatial and temporal concepts and a progressive differentiation precisely as a function of the child's increasing ability to distinguish factors of movements and speed from space and accomplished work. Thus, from a developmental point of view, one is not really in a vicious circle anymore, because in the child, time becomes slowly differentiated from space and is constructed from a coordination of space and speed. Although deriving and abstracted from these two concepts, time continues to bear relationship with them later in our life.

What is important is that psychologically time appears as a resultant, not an intuition. It is a resultant only in the sense that it derives from a progressive coordination of space and speed. This derivation constitutes the foundation of Piaget's interpretations of the development of time.

Let me return for an instant to this quasi-paradox, which in its simplest form comes from the fact that time can be defined as a relation between length and speed, where speed is defined as a relationship between length and time. It is clear that there is no resolution to this paradox as long as one attempts to provide a formal definition of a concept that would be called time but itself arises from a set of nontemporal concepts. The same situation exists for the concept of number, for the logical implication, and for a series of cognitive concepts that are, on an epistemological level, characterized by the same paradoxical aspect and that are, probably not for random reasons, the source of very fruitful problems at the level of the psychology of intelligence.

One can extract oneself from this situation by rejecting the proposition of searching for a definition of the concept of time. The notion that time is in itself a concept is related to an epistemological point of view that one does not need to adhere to. In effect, from this point of view, a

Kantian postulate is not a necessary conceptual prerequisite. What we are left with, then, is the necessity of a structural embodiment of the problem of time.

3. The Structural Approach

The structural approach I outline here bears analogy with the methodology of the structuralists in mathematics, whose results show how much more fruitful it is to look for the structure of arithmetic than for a definition of the concept of number. The essential element of this structural approach to the concept of time is to admit a series, not necessarily linear, of constructions, in which each series corresponds to a sort of "weakened" time (Voyat and Papert, 1967).

a. The structure S_o is a simple structure of order, not necessarily linear and without metrical component. It defines an asymmetrical, transitive relationship of order between events.

From S_o, one can deduce two other structures:

b. Mathematically, the structure S_e has the form $(S_o \times \bar{S}_o)$ where $\bar{S}_o$ is the dual of S_o and $\times$ designates the Cartesian product. Intuitively, S_e is a structure of partial order on the intervals of the structure of order S_o. It allows the comparison of two intervals under the conditions that one of the two intervals is logically enclosed in the other.

In these first two structures, one does not find traces of the paradox resulting from the interdependency of speed and time. In order to proceed to the next structure, one needs only a qualitative, not a quantitative, concept of speed. This leads to the notion of a "generalized watch."

c. One is often tempted to draw an analogy between a watch and a yardstick. This analogy is correct, but one needs to pay attention to the definition of a watch. Should one define it only as an apparatus that generates a movement of constant speed, one would very soon return to the tautology one is attempting to avoid. The watch has, in fact, another property that is theoretically more important than its constant speed: it can be rewound; that is to say, one can act on it in such a way that at a given time, $t(n)$, it finds itself in exactly the same physical state as it was at the moment that preceded $t(n)$. The crucial fact is that one can reproduce, departing from a moment, $t(l)$, a series of events temporally isometric, parallel to another sequence of events that take place at a moment $t(o)$. In other words, one can displace in time or transpose series of events and compare them.

This is the definition of the generalized watch, and one can see that this definition is possible only with the help of at least the structure S_o, that is, with the help of a structure of order. In fact, this generalized watch is possible only to the extent that one accepts the idea that events can be ordered and thus that events are discrete (separable from each other), have an order in their succession, and can thus be put in correspondence.

d. Yet this generalized watch allows us to define a logical relationship of equivalence between separated, temporal intervals. Let us designate by (S_d) the structure thus obtained.

e. Finally, one can deduce that S_d (the relationship of logical equivalence between separated intervals) and S_e (the relationship defining logically enclosed time intervals) generate a new structure in which the relationship of equivalence leads to a relationship of order between all types of intervals: separated, enclosed, or intersecting (S_n).

I have elaborated a construction of physical time that has at least the advantage of demonstrating the role of the factor of speed in two ways: (1) first, it proceeds as far as possible without a structure of speed, and (2) it uses only a qualitative, nonmetric form of this concept.

But what is the developmental meaning of such a structural construction? Let us suppose that a subject possesses the structure S_o (a structure of order). Logically, it would be possible to construct S_e (enclosed time intervals) without the intervention of the factor of speed. Is this psychologically and developmentally possible?

An analysis of some of the Genevan and our experimental results concerning the development of time, at both a perceptual and a conceptual level, provides some answers to this fundamental question. Not surprisingly, the answer shows that although formally possible, it is not psychologically the case.

4. The Piagetian Developmental Approach

Before analyzing these results, let me describe the general form of almost all the Piagetian experiments on the genesis of intelligence. A subject S is faced with a situation that entails a physical apparatus or a verbal presentation or both. In front of this situation, the S is asked to

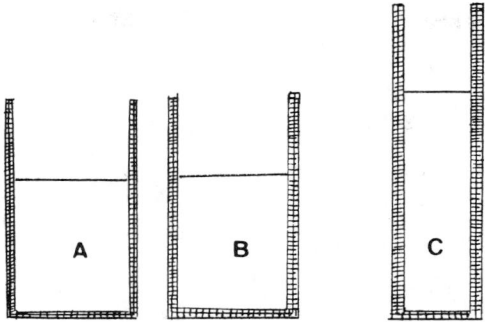

Figure 1.   The classical Piagetian paradigm of conservation of liquids. A = B = C
(from the point of view of quantity of liquids).

solve a problem in finding the actions and/or the manipulations and/or
the verbalizations that satisfy the instructions. The philosophy of the
clinical-developmental method consists in finding a representation of the
S's pathway toward the solution through a series of operations that can
be grossly divided into two types: (a) operations of encoding and
(b) operations of transformations of operations or of composition of
operations.

In a first approximation, one could conceive that the first types do
not go beyond the realm of perception. In a sense, this assumption is
correct, but what is important is less the act of encoding than the selection of information, which leads to the S's decoding.

An example will clarify the problem as well as the mistaken view
of some psychologists who explain the features of preoperational thinking through a simple formula such as, "the child is dominated by perception." Let me address myself specifically to the problem of conservation of liquids (a classical Piagetian situation) poured from one
container to the other, as illustrated in Figure 1.

Let us consider a S who judges as equal the quantities of water in
A and B but considers C as containing more than A after pouring B into
C. It is evident that the S judges quantity according to the level. An
additional proof emerges from the situation pictured in Figure 2, in
which the same S will also think that D contains a different amount
than E. But it is not enough to state that his[2] judgment is dominated

[2] In order to avoid making this kind of writing unduly distracting, I use *he* rather than
he and she and ask for the reader's indulgence.

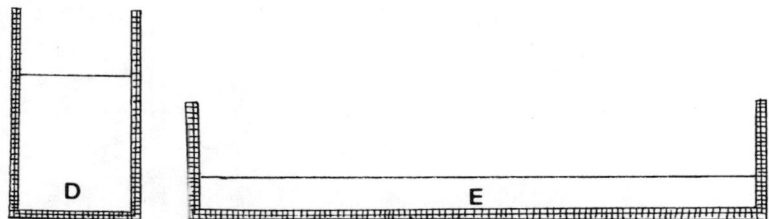

Figure 2.   Immediate possibility of differentiation between the quantity of liquids in D
              and E. D ≠ E (from the point of view of quantity of liquids).

simply by perceptual factors such as the level of D in comparison to E, because in order to believe that D contains more than E, the *S* must also annul or cancel the *global* perception of the liquids to take into account only one factor, namely, the level of water in D. In this action of annulling or canceling a global perception, one has to acknowledge the presence of cognitive factors even in preoperational and perceptually dominated behaviors. But the main point is that the selection of information is always an intrinsic part of an internal representation and therefore necessarily involves cognitive factors.

In effect, in analyzing the general structure of the Piagetian experiments dealing with intelligence, one is seldom in a position in which the dissociation between perceptual and cognitive factors is possible or even justified from the *S*'s point of view. This fact has many implications for the concept and perception of time.

For our purpose, it is important to bear in mind that this state of affairs derives not only from the general form of Piaget's experiments in the realm of intelligence, not only from his insistence on a structural view, but also and probably mainly from the difficulty one has, when representing an internal pathway of thought processes, in differentiating between the encoding and the *S*'s selection of the internally meaningful informations that he assimilates.

5. Presentation of Experiments Dealing with Time

Let me now address myself to a set of specific experiments dealing with the development of the concept of time in 51 normal children from 5 to 10 years of age. Of these *S*s, 18 are between 5.0 and 6.11 years of age (Group I: 5–6); 11 are between 7.0 and 7.11 years of age (Group

II: 7); 11 are between 8.0 and 8.11 years of age (Group III: 8); and 11 are between 9.0 and 9.11 years of age (Group IV: 9).

The material consisted of two boards, each 5 feet in length and 1 foot in width. On each of them two railroad tracks were built on which two engines can move. One is green (V), the other yellow (J) (see Figure 3), and they move according to a constant ratio of speed (V = 2J), which means that the green engine (V) goes twice as fast as the yellow one (J). Each board has two tracks (AB and ACDB), ACDB being exactly twice as long as AB. The same is true for A′B′ and A′C′D′B′, the corresponding symmetrical second track on which J runs. The experimental configuration is represented in Figure 3.

a. The child is first presented with the two engines and asked to designate their colors verbally. He also observes that the two engines can move forward and backward and can take either of the two following tracks: AB or ACDB for the green engine (V), and A′B′ or A′C′D′B′ for the yellow one (J). The child is then asked to designate which pathway he considers as shortest or longest in justifying his answers.

b. The two engines (V and J) move simultaneously: V goes from A to B, whereas J moves through half A′B′. The child, after having observed this situation, is asked to provide a judgment on time: "Did the two engines move for the same length of time? Did one move for a longer or a shorter period of time in comparison to the other?"

c. The experimenter then proceeds to an equalization of the points *and* times of arrival in asking the child to find all the means by which

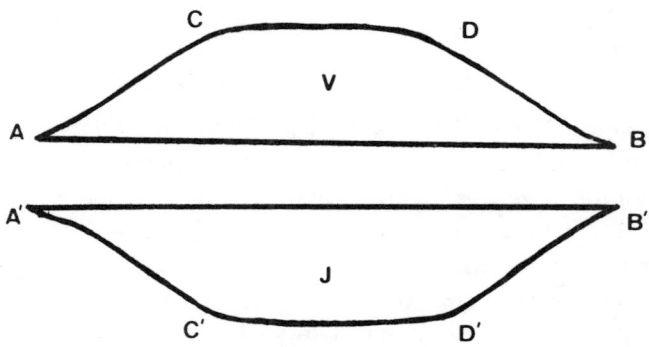

Figure 3.   Experimental apparatus. AB = A′B′; ACDB = A′C′D′B′; ACDB = 2(AB). V, green engine; J, yellow engine.

the two engines could arrive simultaneously at B and B′ and maintain the speed ratio (V = 2J).

There are three possible ways by which the child can comply with these instructions:

1. The child can act on the spatial interval by letting the fast engine move on the longest track (V in ACDB) while the slow engine moves on the shortest track (J in A′B′), V and J both moving simultaneously.

2. The child can proceed through a dissociation between the points of departure, letting the fast engine V depart from A and the slow engine J from the middle of A′B′, V and J both moving simultaneously.

3. The child can separate the time of departure of the engines, which means that V departs when J is in the middle of A′B′.

d. The next part of the experiment is devoted to what is referred to as "enclosed time intervals." The duration of V and J are dissociated from the point of view of time of departure $[Td(1) \neq Td(2)]$ and time of arrival $[Ta(1) = Ta(2)]$. In this case, J (the slow engine) is placed at a quarter of the track A′B′. V departs first from A, and when it arrives at the height of J, J departs and stops at three-quarters of the total length of A′B′, whereas V proceeds to B. The interest of this situation lies in the fact that in this case, judgment of speed is no longer necessary to arrive at a correct judgment of duration because the observation that both times of departures and arrivals are separated is sufficient to provide a correct judgment of duration. In other words, it is sufficient for the child to observe that V departed first and stopped last in order to provide a correct response.

5.1. Comparison of the Length of the Pathways

The comparison of the two tracks (AB and ACDB) has provided unexpected results only in a few of our younger subjects. Almost everyone immediately recognized that ACDB is twice as long as AB, but 30% of our Ss between 5 and 6 considered AB longer than ACDB. This error is interesting, as we will see from the results when these Ss were asked to compensate space by speed, that is to say, when these Ss were asked to act on the material in such a way that the two engines arrived simultaneously at B and B′.

5.2. Judgment of Time

Judgments of time in these situations—in which the two durations were objectively equal, since the two engines departed and stopped simultaneously—provided us with the results found in Table 1.

In (A), the Ss declared that the duration of T1 (V) was shorter than that of T2 (J). These Ss believed that *slower means more time.* In (B), the Ss thought that the duration of T1 (V) was longer than that of T2 (J). These Ss believed that *faster means more time.* (C) refers to wrong answers and (D) to a correct time estimation.

These results are very clear. They show the difficulties that children have in judging two equal durations: only 27% judged correctly at 8 years of age and 33% at 9 years of age, none before. The fact is that the discovery of this equality is neither primitive nor easy, since, for instance, at 9 years of age, 41% of the Ss still thought that the slow engine moved for a longer time.

There is another aspect of our results that is worth mentioning and is crucial: 70% of the children between 5 and 6 years of age judged that the fastest engine took the longest time, whereas this proportion is almost exactly the inverse at 7 years of age, when 77% of the children judged that the slowest engine took the longest time. This spectacular reversal of time estimation is especially important when one inserts these behaviors within the context of preoperational versus concrete operational thinking. From that point of view, the preoperational level of thinking is characterized by a direct relationship between speed and time (more speed implies *more* time), whereas the concrete operational level of thought is characterized by an inverse relationship between speed and time (more speed implies *less* time).

In order to clarify these facts, let us examine the percentages of Ss

Table 1. Time Judgments by Age[a]

Ages	(A)T1 < T2	(B)T1 > T2	(C)	(D)T1 = T2
5–6	30%	70%	100%	0
7	77%	23%	100%	0
8	55%	18%	73%	27%
9	41%	26%	67%	33%

[a] Durations objectively equal.

who admitted or negated the simultaneity of times of departure for V and J (Td1 = Td2) and the simultaneity of times of arrival for V and J (Ta1 = Ta2), as well as the percentages of those for which this simultaneity did or did not imply the equality of the two durations. Here we have five possibilities, expressed in Table 2 by a, b, c, d, and e:

a. (Td1 and Ta1) as well as (Td2 and Ta2) are not judged as simultaneous.
b. (Td1 and Td2) only are judged as simultaneous.
c. (Ta1 and Ta2) only are judged as simultaneous.
d. (Td1 and Ta1) and (Td2 and Ta2) both are judged as simultaneous without implication of an equality of duration (T1 ≠ T2).
e. The correct answer: T1 = T2.

We find the results in Table 2.

Of the total population, 8% of the Ss negated the simultaneity of both the time of arrival and the time of departure. Although 17% of the Ss accepted the simultaneity of the time of departure but not of arrival, only 3% did not accept the simultaneity of the times of arrival (c). This means that the Ss' centration dealt more significantly with the end points. But the most significant observation lies in the fact that more than half of our Ss (57%) admitted the simultaneity of both times of departure and times of arrival without deducing the logical necessity of equal durations.

This last observation is very important: the recognition of simultaneity does not imply *ipso facto* that two simultaneous events are by necessity perceived and/or conceived of as equal in duration. The

Table 2. Breakdown of Time Judgments by Types of Arguments (by Age)

Ages	a	b	c	d	e
5–6	0	33%	11%	56%	0
7	33%	33%	0	34%	0
8	0	0	0	73%	27%
9	0	0	0	67%	33%
Total (average)	8%	17%	3%	57%	15%

Table 3. Justifications of Time Judgments by Type of
Referents (by Age)

Ages	Speed (a)	Space (b)	Time (c)	Coordination of the 3 (d)
5–6	60%	34%	6%	0
7	84%	8%	8%	0
8	55%	18%	27%	0
9	60%	0	0	40%
Total (average)	65%	15%	10%	10%

implication is that for more than half of our Ss, a judgment of equal
duration was different from a judgment of simultaneity of two events.

Furthermore, the justification given by our Ss can be divided into
four types: arguments (a) of speed; (b) of space; (c) of time; or (d) of a
coordination of the three. From this point of view, we obtained the
results in Table 3.

An overwhelming majority of the justifications explicitly referred to
an argument of speed (65%). There was a maximum at 7 years of age
(84%). It is also interesting to note that the largest proportion of the Ss
who used a spatial argument for their justifications were the youngest
children. At 9 years of age, the situation was clear: 60% of the Ss used
an argument of speed and 40% a coordination among the three. This
last behavior is late in its appearance and significantly refers only to the
Ss who recognized the necessity for an equal duration when two events
are simultaneous in terms of their time of arrival and departure. Thus,
it can be inferred that the coordination among space, speed, and time
occurred only for those subjects by whom this implication was actually
recognized. In effect, one can qualitatively distinguish three types of
answers concerning a judgment of time in our Ss.

1. The first type of S believed that the fastest engine took the
   longest time. These children established a direct relationship
   between time and speed and their justifications principally dealt
   with speed.
   S(6.00): "Which one went for the longest moment?" "The

green one." "Why?" "Because it moved faster." "Both engines did not move for the same long time?" "No. Faster means a longer moment."

2. From 7 years of age, another type of answer began to appear that consisted in affirming that the slowest engine took more time because it went slower. This implies the recognition and the discovery of an inverse relationship between speed and time. $S(7.00)$: "Did the two engines go for the same long time?" "No. The yellow moved for the longest time because it moved less fast."

3. From 8 to 9 years of age, a third type of answer appeared in which the S concluded that the two durations were equal after having admitted that both engines had departed and stopped simultaneously.
   $S(9.54)$: "Did they both depart at the same time and stop at the same time?" "Yes." "Did they both move for the same long time?" "Yes." "Why?" "Even if the green one went on a longer path, it went faster for more space and therefore it is the same time for both."

5.3. Equalization of the Points and Times of Arrival

This equalization offers four possibilities to the S. The first one consists in equalizing the speeds of both engines (a). The second possibility can be divided into two parts, each of them dealing with an action upon space: the S can either double the pathway of the fastest engine (b) or modify the points of departure of both engines in such a way that the slowest engine (J) departs from the middle of $A'B'$, whereas the fastest begins in A (c). Finally, the child can separate the times of departure himself, letting the slowest engine start first.

We obtained the results shown in Table 4.

Here again, we observed the primacy of speed as an essential factor in situations in which the task was to compensate speed by space and/ or time and/or points of departure. But there was more. We observed that 30% of our Ss between 5 and 6 years of age gave an incorrect estimation in the comparison of the two tracks (AB with ACDB). An overwhelming majority of these particular Ss (70%) wanted to act on space in this part of this experiment. It was interesting to observe their way of proceeding. All gave a coherent equalization of the points of arrival of the two engines, equalization logically correct (to make the

Table 4. Equalizations of Points and Times of Arrival
(by Age)

Ages	Action on speed (a)	Space (b)	Time of departure (c)	Points of departure (d)
5–6	53%	23%	12%	12%
7	34%	11%	31%	24%
8	38%	22%	22%	18%
9	38%	24%	21%	17%
Total (average)	40%	20%	22%	18%

fastest move on the longest track), except that the compensation was
objectively incorrect. In effect, these Ss wanted the fastest engine (V) to
take the track that they saw as the longest (AB). We have here an
interesting example of logical compensation based on false premises. Yet
it shows quite effectively the child's need for cognitive consistency.

Furthermore, these facts demonstrate a very early possibility of
compensating speed by space. This compensation was made in actions
and could appear as paradoxical when compared with judgments given
on duration, in which at 9 years of age, only 33% of the Ss succeeded. In
this respect, it is interesting to note the instructions we gave our Ss.
When we asked a child to find means by which the engines would arrive
at the same time at the other end of the tracks, we were asking him in
effect to provide a coordination of the points and times of arrival. This
did not imply for him that the two durations were equal. When we
observed here a very interesting example of primitive reasoning by compensation, this reasoning dealt with space, which is an ordinal referent,
and with speed, space being the factor upon which the compensation
was made. Thus, what we observed here was a coordination of space,
not of durations, and it was quite informative to note how the child
could interpret this part of the duality space–speed without its correspondingly affecting his judgment on time.

5.4. Enclosed Intervals

In this last part of the experiment, in which the factor of speed was
not necessary for a correct estimation of the durations because the times
of departure and of arrival were different for the two engines, we

obtained the following results: at 5–6 years of age, 62% succeeded; 60% succeeded at 7 and 8 years; and 73% succeeded at 9 years of age. It is noteworthy that the children could provide correct estimations of time in the case of enclosed time intervals. This means that at all ages they were essentially able to focus solely upon an ordinal referent, the basis of success in this section of our experiment.

5.5. Conclusions

The conclusions one can draw from this set of experiments consist first in a confirmation of Piaget's thesis on a progressive construction of time from kinematic and spatial data. When a child is asked to provide a temporal judgment from a situation in which two durations are both simultaneous and equal but with a different distribution of the ratio between space and speed for the two engines, the child goes through a first stage, in which the ratio between speed and time is conceived of as a direct relationship (faster implies more time), and he then goes through a second stage, in which this ratio is inversed. From 8 years on, and not before, correct estimations begin to emerge. Furthermore, a sizable number of Ss who admit the simultaneity of the times of departure and of arrival do not deduce the necessity for an equality of durations. Finally, in the situation of enclosed intervals, in which the role of the factor of speed is not necessary for a correct response, one observes that time judgments are essentially correct but based upon ordinal referents.

6. Implications for a Developmental Perspective

Piaget has looked at the genesis of the concept of time as essentially based upon a relation between a produced effect (space) and speed. The achieved form of this relation is represented by the formula: ($t =$ distance/speed $= d/s$). Before the child arrives at this operatory coordination, Piaget has observed two stages, which are worth stressing again: in the first one, the child answers that something moves faster and (consequently) it takes more time; in the second one, the child says that it moves faster and (consequently) it takes less time. The essential difference lies in the fact that the first answers take their foundations from the results of the actions; an ordinal referent becomes, in effect, a time referent through its kinematic properties. Faster means further and thus more time. The ordinal referent is at the basis of a time estimation. In this context, the child can also compensate space by speed. It is important to note that in our situation and those described by Piaget there is an interaction of several difficulties for the S. The only satisfactory solution to the problem is to arrive at the equality of durations from a correct interpretation of a simultaneity of times of departure and arrival. Yet, the child can fail to make this implication for two logically distinct reasons: either he does not possess the operation that would lead him from a relationship of order to an operation of time, or he possesses this transitive operation and can use it in some cases but is not able to distinguish those situations in which speed and space are useless in providing a correct judgment of time and thus only the comparison between order of departure and arrival is necessary.

In this second case, one has also to distinguish between two possibilities: the child's inability to eliminate space and speed can be the result of a lack of operatory compensation between these two factors, or it can be the result of the strength and persistence of these factors in the observation of the experimental configuration itself.

Given these distinctions, Piaget's epistemological thesis could lead to a totally different interpretation, provided the two following hypotheses are true:

a. In a situation of enclosed durations, of addition and subtraction of contiguous durations, the S comes precociously to correct operational thinking as long as there is no problem of speed. If such were the case, Piaget would have to acknowledge the development of a set of temporal operations that are independent of factors of speed, and one would interpret the child's answers as symptomatic of the difficulty of coordinating this set of operations with another set of cognitive operations that entail kinetic operations; only a correct coordination of this type would allow the S to provide a correct judgment of time.

b. A second interpretation of Piaget's position would attribute to the operations dealing with speed a formative role in relationship to any operation dealing with time and duration. In this case, speed can also interfere with time, and the development of a coherent and consistent time estimation essentially deals with the progressive coordination of space, speed, and time.

In order to come to a resolution between these two hypotheses, one needs not only to remain in the realm of time as a concept, but one should also deal with its status in the realm of perception of time or subjective time.

7. Perception and Concept of Time

Among the domains of thought, time is probably the entity in which the contrast between formal and intuitive thought is most marked. But even if one ignored all the debate and confusions between these two aspects of what is called time, a language capable of dissociating them would be of great utility. In the formulation of such a language, a number of distinctions would be useful:

a. The first distinction concerns two attitudes, an ontological and an epistemological one. Let us imagine a concrete situation: a subject who has to judge the time it took him to accomplish something may, as is well known, make large errors, varying from 10 minutes to 3 hours for an objective time of 1 hour (Gonseth, 1964). The ontological attitude consists in saying that this erroneous judgment from a physical point of view is nevertheless correct from the point of view of the subject. In this approach, the point is not to know what exists in reality but to extract structural systems of behaviors displayed by children in given situations. In this case, the epistemological view consists in defining what a system of temporal judgment is, in determining if there are one or many systems, and in discovering their logical, psychological, and developmental relationship.

b. Let me return to our concrete experimental situation to make a second distinction. It is possible that a child confronted with the time of a clock after having provided his subjective time estimation might say: "I am wrong; only an hour went by," demonstrating thus that what he intended to do was to estimate physical, objective time. But one could equally conceive of another reaction: "Of course, the clock gives another result; nevertheless, for me the time which went by was ten minutes." If one pursues the inquiry, one finds that this second judgment deals not with time as such but with feelings or sensations of joy, of involvement, of frustration, of well-being, etc.—in effect, a series of emotional and cognitive factors that are summarized by a judgment that takes a temporal form when expressed. I do not deal with the basis of these feelings in this chapter but only with their insertion in a system of judgments about time. From this point of view, I consider only the case of the first subject, the one who attempts to estimate time as the clock measures it. One could even consider this subject as using his own body as a clock: he knows when he begins to be hungry, to be bored, to be tired after a certain time; he can attempt to estimate time according to his physiological reactions. It is evident that this *modus operandi* is approximate and open to gross errors, but a bad clock is nevertheless a clock. What is important is that this subject uses a number of bodily changes and inserts them within a time estimation. This situation is theoretically somehow different when one deals with the estimations of short durations.

Let us imagine an experiment in which the subject is invited to estimate whether the time between a first and a second sound is greater or smaller than between a third and fourth sound. If the time intervals are very long, the subject can use a number of cues, such as his watch or counting, but when the time intervals are very short (a 10th of a second, for instance), the subject can no longer use these types of cues, and in this case, one can properly talk of perception of time. But in speaking of perception of time, we do not want to distinguish between "the perception of duration as such" and "the perception of a rhythm constituted of four sounds," for instance. The essential fact is that the subject's judgment is not made through the use of a composition of cues that have a particular status independent from a temporal estimation.

In this respect, it is interesting to note the results for short durations, between 500 and 4,000 milliseconds (msec) (Voyat, 1963), when Ss were asked to estimate if the last interval was longer or shorter than the first one. The experimental situations consisted of events of flashing light, in which each flash was an event and the time interval between each flash was the interval to be estimated. The salient points of this research on time perception, which used 60 normal adults and 12 children between 7.6 and 8.6 years of age, can be summarized as follows:

a. For intervals of a total duration of 500 msec divided into two or three separate intervals, a general underestimation of the last interval occurred (56%). This underestimation remained for 1000 msec of total duration. For longer durations, longer than 1500 msec, a situation of overestimation of the last interval perceived appeared. This overestimation remained stable and constant.

b. The repetition of the same time perception situation for the same Ss led to a reversal of the illusion. When the S initially overestimated the last interval perceived, he underestimated the last interval perceived after an average of 150 repetitions. The reverse was also true: when a S began with an underestimation, he offered an overestimation of the last interval perceived after a sufficient number of repetitions.

c. An analysis of the ways by which the Ss thought they estimated time show for the adults and the children a systematic reference to an explicit argument of speed and frequencies of events and three complementary modalities of estimation: a first group of Ss felt that they judged the time intervals at random; a second group proceeded to count internally and provided an estimation according to this felt internalized means; and a third group began by counting and proceeded later to give what they felt were random answers. The crucial fact is that there was no significant difference between these three groups.

d. Finally, when the same Ss were asked to provide a judgment on the speed of the events in relationship to each other, we observed a complete parallel and isomorphism in their time estimations. The underestimations of time were in rigorous correspondence with the overestimations of speed for the last interval perceived.

In short, the most significant findings are that even for very short time intervals, duration is apprehended only through a reconstruction that uses an interplay between spatial and kinetic factors. This reconstruction is both intuitive and subjective. Even at this "almost unconscious" level, the S's perception remains consistent with a Newtonian perspective: time and speed are perceived as an inverse relationship, even and significantly for the Ss who believe that their estimations are random.

In the case of perception of time, its apprehension is again the consequence of the interplay of two components: one of accomplished work (the interval between the two events, and thus a spatial referent) and another of generalized speed (the events themselves), where more speed implies less time. The overestimation of the last interval perceived is the result of a centration upon the interval itself, whereas the underestimation is the result of a focus upon the frequencies themselves.

In fact, these results are definite when interpreted within the framework of admitting that this perception corresponds to a construction, not a direct apprehension. This framework is, in fact, almost a necessary conceptual prerequisite for a coherent interpretation of the facts observed. The surprising point is the underestimation that the S provides for the last interval perceived. This underestimation is superficially contrary to the usual laws of perceptual centrations only to the extent that one maintains that our Ss judged time *per se*. It is quite conceivable that our Ss did not perceive time *per se* but proceeded even for these short durations in mental reconstructions using the cues directly available to them: thus, speed and space, not time.

There are a number of our results that point toward the above interpretation. The inverse proportion of judgments of speed and time obtained tend to show that our Ss could in fact very well center upon frequencies from which they deduced time. In this case, the underestimation is no longer contrary to the usual expectation concerning the laws of perceptual centrations; neither is it surprising; the Ss used primarily a cue of speed, of frequencies (a cue that they overestimated when asked to deal with it directly) to estimate time.

The effect of repetition is also important to note. It not only demonstrates the interplay between cognitive and perceptual factors, but it also is easily explainable under the assumption of time "as a reconstruction." Let me take the two distinct cases observed for repetitions:

a. The first case concerns the situations in which the Ss initially underestimated. If one admits that it was the limits of the intervals—the events themselves and thus a cue of speed—that were perceived, then after a number of repetitions, those Ss became capable of shifting their perceptual centration from the limits of the intervals to the intervals themselves. This change of centration provoked a change in their time estimation as a result of an "unconscious awareness," so to speak, of an inverse relationship between time and speed.

b. The second case concerns the situations in which the Ss initially overestimated. Here again, the interpretation requires an interplay between perceptual and cognitive factors in which the Ss perceived the intervals themselves first and then were able to shift their centration from the interval to the limtis of the intervals themselves. In so doing, they came to underestimate the timing of the last interval perceived for similar reasons as in the first case. Perceiving first the limits of the intervals thus leads to perceiving factors of frequencies, of speed.

The pertinent conclusions for our purpose is the essential fact that perception and concept of time are not really different and/or distinct within a S's operational cognitive structure. Both of these aspects of time are intimately related, as is the case in many perceptual phenomena (Piaget, 1969). Time is cognitively and perceptually a unitary concept from the point of view of the *S*. Developmentally, it is the result of a long construction, and psychologically it consists of a reconstruction whose modalities borrow many parallels with Newtonian physical time, particularly at the level of achievement of the concrete-operational period of thought.

8. An Overview of Some Developmental Perspectives

At this juncture, it should be useful to provide a developmental overview of two cognitive tasks of conservation (matter and volume) and their relative age of acquisition in comparison with the development of the concept of time, as presented before (see Table 5).

In examining Table 5, we observe that temporal behaviors are, in effect, of two types: (a) those that have been acquired by a majority of children at 8 years of age, such as the elementary conservations from which conservation of matter was chosen as typical; and (b) those that do not seem to be achieved before 10–11 years of age, that is to say, at the threshold of the formal level of thought.

The discovery of the child of the functional relationship between space and speed is part of the concrete operational period of thought, whereas the correct implicative deduction dealing with a judgment of equal durations is part of the beginning of the formal level of thinking. This distinction is, in effect, the result of the complex nature of the concept of time, which entails many steps in its achievement.

Table 5. Comparison of Ages of Acquisition of Conservation of Matter (I) and Volume (IV) and Time (II, III)

Ages	I[a]	II[b]	III[c]	IV[d]
5–6	16%[e]	16%	0	0
7	32%	33%	0	12%
8	72%	60%	27%	28%
9	92%	75%	33%	30%

[a] Conservation of matter.
[b] Equalization of points of arrival.
[c] Equal durations.
[d] Conservation of volume.
[e] Percentages indicate success in the experiments.

There is a first initial step in which time, as conservation of matter, is a matter of intrapropositional logic. For the child its content is intrinsically part of a spatial and kinetic set. In effect, the child, at this moment of his development, is essentially preoccupied by the elaboration of an inverse relationship between speed and time, whereas at the preoperational level of thought, the child has established a proportional relationship between speed and time. During this period, the child expresses his concept by "more speed means more time" because of his overall focus upon "more," a comparative label implying essentially an order, as an extensive category. Thus, from a psychological point of view, the structure S_e is a primitive functor.

9. The Relationship between the Structural and Psychological Approach

These observations lead me to the questions I initially asked concerning the psychological meaning of our structural formulation. Let us first remember that my construction proceeded as far as possible without the intervention of an argument of speed and dealt with a succession of structures of order and of succession of events. The answer is clear and provides some insights into the child's mind and his way of elaborating the remarkable psychological construction that time is.

First, there is a tendency for the child to remain within the constraints of the ordinal egocentric referent. This aspect is quite general. For the child, an activity or an action takes place in a space that is an immediate referent; things happen in it. For him, space has both reality and concreteness. So has speed, which is almost invariably part of his realm of actions: slow and fast movements are constantly taking place. Yet, both space and speed are also constructed and, according to Piaget (Piaget & Inhelder, 1967), far from simple perceptual encoding.

In the realm of time, the first abstract act that the operational child achieves is to invert the relationship between speed and time. This inverse relationship may very well derive from his experience; he has many times had the opportunity to see space and speed interact. Time as an order of succession is already understood at this age. This understanding leads the child to the concept that events have not only an order or a succession but also a speed related to them. Thus, it is not surprising that the abstract inverse relationship between speed and time is mastered at a time when the child comes also to understand concrete invariants such as conservation of matter. It is interesting to note that this invariant is in fact implied in the structure S_e, which allows for the comparison of sequences of events and their correspondence.

Furthermore, the structures S_d and S_e are late in their psychological occurrence. They coincide with the emergence of the initial step of the formal level of thought. This coincidence is again developmentally quite understandable, and our observations concerning the concept and the perception of time point to the following direction: conservation of volume and the implicative relationship "simultaneous times of departure and of arrival imply equal durations" both entail a new type of abstract coordination. The simultaneous operatory coordination consists, in the case of the concept of time, of maintaining constant two different sets of operations. One deals with the apprehension of time of departure and of arrival as a total event and compares it with another identical event; the second deals with the comprehension that this simultaneity implies equal duration. This coordination is formal in nature, and it is not surprising that only in adolescence does the child come to understand its abstract meaning.

What can be inferred about the cognitive nature of the reality of time from an epistemological point of view? In order to answer this question, one might ask if the preoperational child has a Newtonian or an Einsteinian sense of time. The answer is not immediate; on the one hand, one is tempted to say that his sense is quite relative, since there is no initial recognition of simultaneity. The first order to impose itself on the child is an order of succession of events. This is also a spatial referent, which becomes temporal only later. In this respect it is interesting to note Abelé and Malvaux's (1954) attempts in France. Using Piaget's findings, they reformulated from developmental data the concepts of speed and time in a relativistic perspective.

According to Piaget (1970), the concept of time is "an intellectual construction. It is a relationship between an action—something which gets done—and the speed with which it is done" (p. 70). This construction is based on operations that are similar to those embodied in logicomathematical thinking. First, there are the operations of seriation, of ordering events in time: B comes after A, C comes after B, etc. Second, there are operations that are similar to the operations of class inclusions: if event B follows event A and event C follows event B, then we conclude operationally that time interval AC is longer than time interval AB. This operation corresponds to the logic of classes, to the concept that the whole is greater than the parts. Finally, we have the operations of measurement of time, which are a synthesis of these two kinds of operations, implying a synthesis between ordering and classifying. Again, structurally there are good reasons to observe that time is in its final stage a concept that is inherently part of the formal level of thought.

As an intellectual construction, time is, as a preoperational concept, part of the objects, part of the events, part of the actions, part of the movements. Time is in this sense both projected upon and internalized from objects. Therefore, it seems essentially to be relativistic. As a concrete concept, time is clearly Newtonian. It is only at the formal level of thought that the adolescent is able to return deductively to its relativistic meaning. Yet between these two stages of development, there is a fundamental difference; at the preoperational level of thinking time is not internalized as a distinct concept. Therefore, the child has an awareness of time only in relationship with objects, events, and persons, as embodied in changes and assimilated to the child's own actions. This aspect is the basic referent, not the awareness of time *per se*.

At the formal level of thought, the hypothetico-deductive form of reasoning—the combinational nature of this structure, the operations upon operations that this stage prompts—implies a further level of abstraction that allows a new relationship between the concrete and the abstract. This new dialectic, concrete and abstract in nature, allows both a level of equilibrium and a further development, all embodied in this formal reasoning. It is not surprising, then, that time, as one of our very abstract constructions, has the same epistemological status as those other abstractions whose values are to maintain our mental, adaptive life.

References

Abelé, J., & Malvaux, P. *Vitesse et univers relativiste*. Paris: Sédès, 1954.

Fraisse, P. *Psychologie du temps*. Paris: Presses Universitaires de France, 1957.

Gonseth, F. *Le problème du temps*. Neuchâtel (Suisse): Editions du Griffon, 1964.

Janet, P. *L'Evolution de la mémoire et du temps*. Paris: A. Chahine Editeurs, 1928.

Newton, I. *Philosophiae naturalis principia mathematica* (*Natural philosophy of mathematical principles*). Geneva, 1739. [A. Motte (Trans.), F. Cajori (Ed.). Berkeley: University of California Press, 1946.]

Piaget, J. *The mechanisms of perception.* New York: Basic Books, 1969.

Piaget, J. *Genetic epistemology.* New York: Columbia University Press, 1970.

Piaget, J. *The child's conception of time.* New York: Ballantine Books, 1971.

Piaget, J., & Inhelder, B. *The child's conception of space.* New York: Norton, 1967.

Sivadjian, J. *Le temps: Etude philosophique, physiologique et psychologique. Vol. 1. Le problème métaphysique.* Paris: Hermann & Co., 1938.

Voyat, G. La perception du temps: Etude d'estimations de durées brèves. *Archives de Psychologie,* 1963, *39,* 153–154. (Genève)

Voyat, G., & Papert, S. *Perception et notion du temps.* Paris: Presses Universitaires de France, 1967.

Editors' Introduction

If, as the authors of this volume believe, time is a highly subjective phenomenon, what are the means by which we represent time? Is time truly linear and constantly advancing in an irreversible progression? Is time spatial, so that we can think of "regions" or "surfaces" of the past, the present, and the future that sometimes touch, sometimes merge, and sometimes are sharply delineated by rigid boundaries? Can time be represented as points or "atoms" of experience, each one an isolated packet or nugget of experience? Thomas Cottle shows that all of these representations have been extensively discussed and debated in 19th- and 20th-century philosophy and phenomenological psychology. More important, however, is the fact that people spontaneously use these modes of time representation in speaking of meaningful events in their own lives.

Cottle shows the uses and roles of different modes of time representation reflected in interviews in which adolescents and young adults discussed their experiences of the personal past, present, and future. By examining time within the context of each person's life, Cottle brings about a rapprochement among linear, spatial, and atomistic views of time. He shows that people do not exclusively use any one mode of representation but rather that they can employ each mode as situations and needs arise. Each temporal mode frames experiences in quite different ways and allows individuals to construct the essential meanings of their lives.

The Time of Youth

Thomas J. Cottle

1. Memories of the Past

The issue of studying time turns on a challenging question: Is there really a phenomenon of time that exists apart from any individual, or does the concept of time reside only in one's perceptions of it? Does time possess objective and subjective features, or do human beings bring what they feel to be objective and subjective evaluations to their understanding of time? Aware of these complex questions, Frederich Kummel (1966) wrote

> The problem concerning the nature of time cannot be separated from the many forms of its interpretation in the history of thought . . . for the very character of our contemporary conception of time is but the result of a long intellectual development. . . . Time, having no reality apart from the medium of human experience and thought, is relevant only in its various historical interpretations (p. 31).

In beginning our discussion, we must keep in mind that in the course of history, people have devised methods of dealing with time that in effect render time objective and measurable. A clock and a calendar are devices that measure the passage of chronological time. Thus, while two people may disagree over whether an hour is a short or a long period of time, they would agree that an hour contains 60 minutes and that it can be divided into smaller units of time called seconds.

Thomas J. Cottle • Children's Defense Fund of the Washington Research Project, Cambridge, Massachusetts.

But the concept of time also possesses subjective properties. We might ask a man to define the duration of the present. Let us even supply him with clock and calendar units—minutes and hours, days, months, and years—to help him with his answer. Again, he will recognize that even with these so-called objective units of time, he cannot arrive at a purely objective definition of the present's duration. He must offer a subjective definition, a report of his feelings about the duration of the present.

The tension between objective and subjective responses to the meaning of time is an important one. It raises the question: How does one perceive the passage of time? In philosophical terms, does one perceive time in a *linear* or in a *spatial* framework, the framework Henri Bergson called a *durational* framework? Because this distinction is of the utmost importance to us, we take a moment to explore it.

As much as any other philosopher, Bergson believed that the passage of life was a continuous one. Because to live is to grow old, time reveals itself partly in the way one experiences one's inner life. "This inner life," Bergson (1964) wrote, "may be compared to the unrolling of a coil, for there is no living being who does not feel himself coming gradually to the end of his role" (p. 140). Bergson was quick to assert, however, that no moment can possibly be identical to any other moment; there is always one's memory that differentiates one moment from another. I know more this minute than I did a minute ago and less than I will know a minute from now. Thus, not even two consecutive seconds can be identical. The earlier second lacks a memory of itself, which the later second will retain. As Bergson (1964) said, "A consciousness which could experience two identical moments would be a consciousness without memory" (p. 140).

The following passage is taken from a conversation with a woman in her early 20s, the mother of two children:

"Now my mother, she couldn't have been more unlike my father. First off, she was a whole lot smaller than he was, but she was strong. Lots of times I felt her arms and they were hard, like iron rods. She worked all her life too, field work, same as he did. Put in the same ten hours like him, only she didn't have no rest on Sundays. I can close my eyes and see my mother doing everything around that farm, milking, hauling, turning over the dirt with this red pitchfork my daddy painted special for her. He gave it to her, I remember. Told her there wasn't any sense him getting her ladies things 'cause they'd never see the day when she'd get to wearing them. But I saw her in all those poses, and never once do I remember that woman sitting down to rest. She was out of her bed, washed and dressed before any of the children got up, and never went to bed before us, even when we were grown up. Never even saw her lie down for a minute to take a break. I'll bet the first time she ever lay down to rest when the sun was shining was the last days in the hospital, right before she died. I remember my daddy saying to her how she looked funny lying down in the middle of the day. It was true, too. His day started at four A.M. Hers did too. Stayed that way for both of them up until the time they died. Everybody said she'd way outlive him the way he worked, but she didn't. It wasn't no surprise that he followed her three years later.

"They were good people. He was tough, putting down the law, and she was soft. When you wanted soft, you went to her. When you needed something, like money or to get something material, you went to him. Made it easy for us. When one of the children would have some kind of problem, it was pretty easy to know who to talk to. You better know, too, that we behaved. Maybe once we'd get out of hearing distance out beyond the road, we'd scream to wake up the devil. But when Mother and Daddy could hear, or when they'd be walking with us, like to church on Sundays, then you saw the best-behaved children in all of Arkansas.

"I keep telling my children they'll want to know about my past someday. 'You don't have to know nothing about history now, but you'll be begging me one of these days.' They say they don't have much interest in Arkansas farms. I don't much blame them, although it's always seemed to me that a person ought to know something about his roots. There's a connection, I tell them, between my daddy working a field near Helena and your daddy working for the phone company here, but they don't see it. So damn much has changed, it's hard to believe all of it's happened in just one lifetime. I can't believe that my life could be so different from my mother's. It's not that many years, and here I am with a dishwasher and TV. Hell, we got a two-car garage. Only got one car, but we're all ready if we could ever afford another. My father didn't even have a shed for his machinery and tools. Used to lean 'em up against the side of the house and cover 'em with a big hunk of canvas. And here we are sitting with a two-car garage. If you're looking down at us, Daddy, be sure you know we got room for you.

"It's hard to understand. I guess one generation makes all that different. But see, even though only a few years have gone by, changes are happening all the time. It doesn't go slow anymore like it used to. Ten years now is like a lifetime when my grandparents were alive, half a lifetime when my parents were alive. Nobody can afford to wait no more. You wait and you miss out on too much. Times change, and places change. Another thing that's different is the way we plan for our children, teach them to think about their futures. Where I came from it was enough just to hope your children wouldn't have to live the way you did the rest of their lives. No one was going anywhere in the society, but some folks could arrange to have their children get out. See what I'm saying? But that's not true anymore. Least for folks like us. We're making our children think about careers. When I grew up there was only the day you were living to be thinking about. There was no use wondering about tomorrow. It'd be the same or you'd be gone or dead. Your job was to get through the days, one at a time. My husband and me, we get ourselves and our children through the days, the weeks, the months, the years, too. And the reason we're doing it this way's so that *our* children can spend all their time getting ready for the future. We'll carry 'em as long as we can, don't you see. As long as they need us. When I was a child, nobody carried nothing for nobody. You barely had time to carry the babies. More likely you'd be setting them down somewhere and moving on with your chores. Lots of time now. Years go by quick, but everyday I know I can make just a little bit of time for myself. So like I say, we're moving out and up a little, and my children are going to be moving up even higher.

"Every time I talk this way though, I get a funny idea into my head. Here we are, living so far beyond our parents' means, breaking our necks in old America so that we might live a little easier and make it all right for our children. But every once in a while I get this idea that my children are going to come to me someday with their bags all packed, dressed up real nice, you know, and say that they've decided what they're going to do in the future. 'What's that?' I ask, thinking they're about to let me in on some beautiful surprise. 'Well,' they say, 'we've decided what we want to do is go to Arkansas and work on a little farm. Ain't so good to be even half way rich,' they'll say. 'Oh, my God Almighty children,' I'd tell them, 'you must be crazy.' Course, my mother and daddy weren't crazy. Even without going to school much, they were good folks. I was about to say they were special folks, most special I suppose I've ever had the good fortune to know."

Throughout his work on the meaning of time, Bergson advocated the notion that time flows as a continuous uninterrupted line. In this way, he associated himself with many physicists, particularly those who opposed the notion that time was periodic and interruptable or that it could even be slowed down. Simultaneously, he disassociated himself from philosophers like Spinoza and Descartes who had argued that the passage of time should be conceived of as atomic. According to the physicist G. J. Whitrow (1961)

> Descartes was compelled to postulate that the instants at which creaturely beings exist must be discontinuous or atomic. Temporal existence must, therefore, be like a line composed of separated dots, a repeated alternation of the state of being and the state of non-being (p. 156).

By postulating his theory of the continuous flow of time, Bergson was not only opposing the notion of periodicity in the flow of time, he was differentiating "temporality," or the continuous, indivisible flow of time, from "spatiality," the conception of time as atomic and divisible. Bergson would have agreed with Van der Leeuw, who warned that when one reduces a melody—time in this case—to its individual notes, there is no longer a melody. Music requires a continuum of notes for its total effect, although by reducing music to its elementary notation, one is still able to understand it (see Van der Leeuw, 1957).

Bergson's notion of time as linear in configuration—derived exclusively from the perceptions of one's inner self—strongly opposed what Immanuel Kant had written. Kant had long argued that the intuition of time and the process of representing time and change to ourselves were found essentially in spatial terms:

> For in order to make even internal change thinkable we must represent time (the form of inner sense) figuratively as a line, and the internal change by the drawing of this line (motion) and so we are obliged to employ external perception in order to represent the successive existence of ourselves in different states (Kant, 1934, Chapter ii, Section 4).

Here then, we have an argument for examining the bearing of external factors on the way we sense time internally. But in Kant there is a second important aspect of time perception. Notice his phrase, "to represent the successive existence of ourselves in *different* states" (italics added). Implicit in this phrase is the importance of memory, the recollection of prior moments that, in the linear conception of time, can no longer be experienced.

It is clear from his viewpoint that Bergson would emphasize in his philosophy the overriding importance of memory. Present experiences, he alleged, are inevitably tied to the immediate past, to what the individual has just experienced. Indeed, the present derives meaning from the past. Whitrow (1961) expressed this same idea in the following way: "Practically we perceive only the past, the pure present being the invisible progress of the past gnawing into the future" (p. 87).[1]

What Kant's statement suggests, however, is that memory is experienced in different ways. We can say that the minute that has just ended can never be experienced in the same way again. Although we can recall the minute, it is not the same as experiencing the minute as we experienced it originally. This is the essence of the concept of temporal succession: the past cannot be regained.

A 19-year-old woman reported the following dream. She approaches the door of her own home. She hears noises and is pleased that someone is inside. She has been afraid that no one will be there to welcome her. She knocks lightly and the door, which has been unlatched, opens. As no one is in the front hall, she walks with some trepidation to the living room. Sitting together on the couch are her mother and father. They are leafing through the pages of a photo album. The scene of her parents together makes the young woman feel happy. They look extremely young, the age when they were married. Her father's skin is smooth, his rich black hair is neatly combed. Her mother is wearing a flowered print dress that makes her look like a little girl. Her hair is tied with silk ribbons. The house is small and the young woman knows there are no bedrooms. There are no children in this house.

The young woman clears her throat hoping to draw her parents' attention, but they do not hear her. "I'm home," she says, but they do not respond. She rattles objects on the table in front of the couch, but

---

[1] This position of Bergson's was attacked by Russell, who claimed, according to Whitrow, that in an obviously circular approach, Bergson had in effect unwittingly removed time from his argument. Ernst Cassirer (1957) also reflected on this important aspect of Bergsonian philosophy when he wrote, "Action is determined by the historical consciousness, through recollection of the past, but on the other hand, truly historical memory first grows from forces that reach forward into the future and give it form" (p. 188).

still they do not react. Finally, she sits in the tall easy chair across from the couch and stares at her parents. They seem so happy. She begins to sob. "I'm sorry I returned. I'm sorry I'm home." Her parents continue to enjoy the photo album together. The young woman stands up and runs for the front door. As she passes through the front hall, a group of her parents' friends, the ones whose pictures appear in the photo album, enter the house. A party is about to begin. Hearing the sounds of their guests, the young woman's parents put the album on the table and rise to greet their friends. The young woman watches everyone kissing and shaking hands. The young woman's mother disappears into the kitchen and returns with a tray of food and cold drinks.

When everyone has had something to eat, the young woman's mother whispers something to her husband. He nods excitedly. She walks into the front hall and reaches the bottom of the stairs. Looking up toward the second floor, she calls the young woman's name. "Come downstairs, we want you to sing a song for the company. Everyone wants to hear you and we do too. Please come down."

"But I'm right here," the young woman says to her mother. "Look, it's me, I'm here. I'm not a little girl anymore."

Her mother continues calling upstairs. "Come on, darling, you don't need to be shy. You're all dressed in your lovely party dress and party shoes, and we've practiced your song. You'll be wonderful. Everybody wants so much to hear you sing. Really they do."

"Mother, for God's sake," the young woman shouts out. "Those days are gone. I'm grown up. So are you, and so is he," she says pointing to her father, who is now helping his wife coax their little girl downstairs. "The past is gone forever. Haven't you heard? We're grown up!"

"Come on, love," her father is calling. "We'd consider it a special occasion if you'd sing. Sing just for Daddy. Okay? Forget all about the rest of the people. Just sing for Daddy. Mommy will play the piano, and you'll sit on Daddy's lap."

From the landing on the second floor, the young woman hears the sound of a child descending the stairs.

"It's past," the young woman pleads with her parents. "None of this exists anymore. Doesn't anybody care what I'm talking about? This is years ago, and now it's now, the present. What's wrong with everybody?"

The young woman sees her parents reaching their arms up toward the landing, expressions of joy on their faces. She glances up and sees the feet of a small child coming slowly down the stairs. The child is wearing white socks with little flowers on them and shiny black party shoes.

The young woman runs screaming and crying from the house.

For other philosophers, Bergson's reasoning did not appear to capture the way in which time seems to be experienced by human beings. Kummel (1966) argued that "succession alone cannot explain the complete essence of time" (p. 38). A feeling we have about time may be more useful and important to us than the fact that we cannot actually recapture past moments. It does seem that the act of memory allows us to call back prior instants, and while surely the recalled instants are not identical to the instants as they were originally experienced, the fact that they are lodged in our memory means that they are not gone forever. We can recapture the past, albeit incompletely, through memory.

This point of view, arguing that memory can recapture the past, is an essential aspect of the so-called spatial or durational conception of time. The conception is called spatial because it suggests that instead of conceiving of time as an unbreakable chain of instants, we can use our imagination to lift a past instant out of its place on the continuum of time and drop it into another place. In the case of memory, we retrieve past experiences and relocate them in the present. As some philosophers put it, we conceive of past and present instants as correlative rather than linear and successive.

To repeat, the phenomenon of memory is not only important for an individual's perception of time, it is an important part of the scheme that philosophers and psychologists have developed in their attempt to represent the meaning of time. For what is temporality other than the perception of events that are bound to the present but that must inevitably be replaced by still newer perceptions?

From the reflections of an 18-year-old woman:

"A mixture of messages destroyed me. It was the urge to move away and be independent when my mother wanted me to be the baby of the family forever. She was afraid of my maturity, afraid that I would surpass her. She wanted me to be spared pain, even though she knew that no one can prevent a child from knowing pain. The greatest pain is not knowing it directly but realizing it is there everyday in one's house.

The greatest hurt of all was my mother's inability to accept me for what I was, and what I became each day.

"She never accepted that people change. Aging crippled her. She despised the fact that I was the young one *she* wanted to be, the single and free one. She tried to push herself backward in time by living through me, or by making me into something that pleased her. Her way caused me to skip over time, to jump ahead by years. She dreamed of a life where people age in one direction until they reach a point when they begin to become young again, demature. Then, when they reach a certain demature age, they begin to age again in the other direction. Then they demature. Back and forth, maturing, dematuring, until the cycles become narrower and narrower. Finally, there is no more time in which to age, no more time in which to become youthful, and people are locked into an age that God has determined for them. They lead the rest of their lives at that age, no matter how old they are, no matter how old they choose to be. They live with a secret age that only *they* know about and are locked into. For my mother, that age was set by my father's leaving. If she kept me young, she kept herself young. She kept life away and death away.

"She tried to make everything still, and she fooled herself into believing that she could get away with it. She was a robber, hoping that if she held out long enough, the statute of limitations would pass and she would be free of blame, free of guilt as well. But she was a fool. She should neither have hid from the law nor turned herself in. She should have realized that life goes on and that one makes decisions and lives by the results of those decisions. No one must determine your life but yourself. If you have no plan, no goal, then at least you can fall asleep at night and stimulate yourself with dreams. And when dreams begin to seem insufficient, then you can dream during the day, compose life poems that no one can destroy. Even if you cannot remember them, you can pride yourself that you had the imagination to create them."

2. Expectations of the Future

If memory of the past forms one basis for perceiving time, then expectation of the future forms another. For many philosophers, the concept of the future is inferred from experiences in the past. Recognizing that we first experience each moment in the present and then watch it retreat into the past, we begin to realize there is something called the future, a period of time in which moments that have not yet occurred exist.

Notice again how the distinction between linear and spatial conceptions affect how one perceives the future. In the linear conception, the successive nature of moments makes it impossible for us to know the future. Accordingly, we must await the entry of future moments into the present in order to experience them. In contrast, there is the spatial conception of time. Here we correlate past, present, and future moments, while recognizing that time is linear and that people do become preoccupied with expectations of the future, impressions of the future, or what Husserl called "protention." While expectation derives to some extent from memory, expectation is not identical to memory. We can recall expectations we have had, but when we recall them, they are no longer expectations but *memories* of expectations.[2]

Because expectations, recollections, and sensations of events occurring in the present do come into consciousness, we can demonstrate what has been called the spatial or durational conception of time. That we are involved with the past, the present, and the future; that we believe them to be related in some way; and that we live with a feeling that there is more to time than the linear passage of succeeding moments, all of these constitute the basis of the spatial conception of time.

In large measure, we cannot experience the phenomenon of time until it is spatialized, until we detime it. We merely infer what time is, even while we experience it. Perhaps this is the reason Emerson called time "a poison." The line of guests arriving at a party and our impressions of these guests, as well as our recollections and expectations of them, reveal the fundamental tension between so-called objective or linear conceptions of time and the subjective or durational or spatial conceptions of it. This is a crucial distinction we must make. In fact, it turns out to be a distinction that all people make in thinking about the meaning of time. But let us again remind ourselves that both concep-

---

[2] It is interesting that in his theory of protentions and retentions, Husserl seemed to allege that perception of the temporal event does more than merely activate processes of the present. If each act of perception is composed of endless protentions and retentions, then will not life experiences seem endless in terms of anticipations? Moreover, has he not thereby rendered being-in-time infinite?

tions of time, the linear and the spatial, are relevant to us as we contemplate the meaning of time. In other words, we bring temporal and spatial criteria to bear in our perceptions of time. To quote Kummel (1966):

> Time is traditionally described as a fragmentation of successive vanishing moments; one can, however, just as logically assert the integrity of time based on the inner correlation and coexistence of its parts. Only the two definitions taken together can fully describe the nature of time (p. 45).

It is essential, therefore, that we not think the two conceptions are antithetical. Instead, we must recognize that the spatial conception of time relies heavily upon the linear conception. The spatial conception does not imply that past, present, and future instants are mixed up with one another, or even that they occur simultaneously. There is always a period of time called the future that we have not yet experienced and a period of time called the past that we have already experienced. The two periods are not identical. In a paradoxical sense, therefore, the spatial conception of time rests on the linear conception, the conception advocating the succession of instants.[3]

In reviewing Bergson's theory of consciousness, Alfred Schutz (1962) wrote:

> *Attention à la vie*, attention to life, is, therefore, the basic regulative principle of our conscious life. It defines the realms of our world which are relevant to us; it articulates our continuously flowing stream of thought; it determines the span and function of our memory; it makes us—in our language—either live within our present experiences, directed toward their objects, or turn back in a reflective attitude to our past experiences and ask for their meaning. (pp. 212–213)

This concept of an attention to life helps us to understand what is happening as people order experiences.[4] It is Schutz's contention that reflecting on the past and anticipating or expecting the future, although dif-

---

[3] Kummel (1966) summarized this point in the following passage: "Thus arises the apparent paradox that the amalgamation of time periods into a mere succession inevitably results in their mutual exclusion, whereas their distinction makes possible their harmony within an articulated unitary structural interrelation" (p. 44).

[4] Once again, it could be argued that the location of experiences in time cannot truly yield temporal orientations since the acts of listing and locating experiences are themselves acts of the present only. Reflecting on past experiences does not carry us back to the past, nor does expectation thrust us forward to the future. As Heidegger said, past and future belong in and to the present, for both are created and filled by activities of the present-memory, anticipation, etc.

ferent activities of the mind, encourage us to conceive of time in spatial
rather than linear terms. Reflection revives or recapitulates experiences
to the extent that one becomes *reinvolved* with the past, whereas expectation serves as a *rehearsal* for future action and thereby gives one the
feeling that he or she has some sense of what the future will be like.

These notions become a bit more complicated when we consider
John Dewey's idea that the future remains forever *empty,* because it can
be filled only with fantasy, while by definition the past includes previously enacted experiences that are retrievable through memory (see
Dewey, 1957). A so-called future orientation is a predisposition in the
present to prepare or rehearse action that may help us to make an
empty period (the future) seem filled. A past orientation, on the other
hand, because it implies preoccupation with completed action, helps us
little in preparing for the future.

Given the notions advanced by Schutz and Dewey, we now ask,
does a future orientation imply a desire to fill a period of time that by
definition must always be empty? Does planning for the future or anticipating it give persons the feeling that they are filling the future, thus
making it seem less ambiguous, less threatening? Does a future orientation, moreover, imply a belief in one's capacity to shape the future? Or
instead, might a future orientation imply a dissatisfaction with the past
and thereby a devaluing of memory and reflection?

Theoretically, the type of future orientation resulting from a
genuine *attraction* to the future, should be different from a future
orientation resulting from a desire to *avoid* the past and present. What a
person is attracted by or seeking to avoid should also influence the
person's perceptions of the past, the present, and the future.

Let us take these notions and consider them in the context of two
forms of reasoning, the *atomistic* and *Gestaltist.* Although these two
terms have been defined by many writers, one in particular, Victor
Gioscia (1966), has defined them explicitly in temporal terms:

> Someone who is accustomed to thinking spatially one part "at a time" is
> referred to in philosophical language as an "atomist." He is a man who
> thinks that things consist of a sum total of component parts. . . . He is
> replaced . . . by the fellow who also thinks in a fundamentally spatial way:
> He visualizes but takes the entire configuration all at once. We call him a
> "gestaltist"; things for him do not consist of a sum total of their parts; a
> thing is a thing, a molecule is a molecule. One can break it up into its
> component atoms but while one has it, it's a molecule. (p. 3)

Holding Gioscia's words in mind, let us attempt to describe Gestaltist and atomistic conceptions of time.

The Gestaltist notion of time suggests that the present passes continuously into the future, just as the past flowed continuously into the present. The atomistic notion of time suggests that the time zones are distinct; that is, they are separated from one another. Apparently, people hold atomistic and Gestaltist perceptions of time simultaneously in the same way that they hold linear and spatial conceptions of time simultaneously. If this is true, then one's sense of causality, which develops in great measure from conceptions of time, is affected by these fundamental styles of thought: persons who tend to see parts within a particular whole as being disconnected should be less capable of perceiving relationships between these parts than persons whose style of thinking urges them to find relationships between parts. The Gestaltist perceiver, therefore, should see not only relationships between time zones but ways in which events in one time zone affect events in other time zones.

Having said this, we now ask, which develops first, a style of reasoning, be it atomistic or Gestaltist, that influences one's perceptions of time, or one's perceptions of time, which then help to shape one's styles of thought? Although we do not possess sufficient material here to answer this question fully, we might wonder about the perception of time as being discontinuous in light of what Bergson meant by the spatial conception of time. If, for example, the temporal horizon is seen as discontinuous, is it possible for people to believe in their capacity to affect future outcomes (*cf.* White, 1959)?[5] Furthermore, does such a perception affect people's conceptions of their own destiny? For that matter, can one understand the concept of destiny if one perceives the temporal horizon as being atomistic?

Eddy Kelly, a suburban boy of 15: "I think a lot about my future. These days you have to think about it. There's so much competition that you can wreck your chances when you're as young as I am. I've already sent away for college catalogs. Last week, a group of us went into the library to work on term papers for English. I was looking around in their shelves and I saw they had *all* the catalogs. That's all I read for three hours.

[5] One wonders, too, whether perceiving the horizon as atomistic might yield a sense of feeling alienated from present events and experiences.

"I'm in a pretty good position out here. Everybody expects you to do good. I mean, they're pretty shocked when kids like me don't do well, don't go to college. Like all these kids who take drugs? That surprises a lot of people. Everybody reads in the papers about the suburbs, people getting divorced, and all the rich kids on drugs, but at school, everybody's expecting us to be smart.

"Must sound strange hearing me talk like this, huh? Some guys I know, they're taking their life easy. They just go along not letting anybody see what's bothering them. But I talk to my folks more than a lot of guys do. My dad's really taught me a lot. Like, how a man's really got to work, and if he does, how the rewards come. I saw where he grew up. You can't believe a guy could accomplish so much, and what the heck, he's only forty something now. He could still get a whole lot more. 'Course, it's not the same for me since I'm starting off way better than him. But still, I'd hate to see my life be nothing more than throwing away lots of good chances. Some guys don't care. But I'm not going to be any failure. Too many people are counting on me. Not only in my family, but at school, too. Later on in life, like my dad always says, there are people waiting around, watching you, hoping maybe that you'll fail. But right now, it's like this whole town is out to make life good for us.

"We were talking yesterday in social studies. Mr. Hamblin told us to imagine how our lives would be five, ten years from now. We wrote these papers for him and then talked about them. Everybody was a success. College and jobs, and people making money and getting married and taking trips. We really sounded like big shots. But then we talked about competition for the *best* schools and the *best* jobs, and everybody started to laugh a little, you know. But the interesting thing was how all that competing came about because we were born pretty well off. Those kids who go to school in the city, a lot of them are going to compete with each other the same way as us, but the chances just aren't there. Some of those kids, even the ones who do the best, won't be as well off in the end as some of us who do lousy. Boy, if that's the way it worked in sports, trying out for a team and making it but still, you know, not making it, I'd be just as angry as those kids probably are."

Mitchell Walker, a working-class boy of 15: "I can't always be sure what they're saying when they tell me to stop worrying about the future. Maybe it's good advice. If you worry too much about the future, you won't get tonight's homework done. But sometimes I think they're trying to tell me to forget it. I think what they'd like to say is that the old Walker kid doesn't have the smarts to make it. I got the smarts all right, the question is, is there a college I can get into? I could always go to some junior college, or a community college somewhere, but from what I hear you kind of pass through those places and come out the other end looking like you did when you went in. And anyway, there's always the money problem.

"What happens, say, if I do get in and we can't get the money to let me go? I'd have to keep working 'cause it isn't going to get any better at home. My father's pretty sick already. He'll be earning a whole lot less by the time I'd be ready to go. So maybe if I went to college, it would only make him feel worse. You know what I mean? Maybe he'd feel pressured to work harder so I could get through, and maybe that extra work would kill him. Vocational school is what this school wants for us. The brains go to college and we're supposed to learn how to fix cars. And the girls go to cooking and nursing classes. Everytime someone like me wants to go to college, it's like it messes up their plans. When we get older, they'll tell us don't apply to college, but some of us do anyway. The teachers are happy if we just pass. They don't want anything for us.

"Then on the other end, you got all those fancy kids going to school in the suburbs. They can't fail. There's no way their school's going to let them fail. If they have to, they'll drag every kid out there into college even if the kid doesn't want to go. 'I don't want to go. I don't want to go,' the kid will be crying. 'You're going, kid!' the school will say. 'You're going. What do you think all this money we've been spending on you has been for?'

"Rich kids got their problems. They got pressures on them, but they don't know how good they got it. When one thing falls down, they've got something else. If their family falls apart, they got their school. If they do bad in school, they got their friends. If their friends leave them, well, at least the homes are pretty. Here, when things go they really go. You start out bad and end up bad. If your family breaks up, all you got are lousy schools and no place to go where you can get yourself together. It's like we're in jail over there. Classrooms are jail, cafeterias are, too. And the halls, man, they really are jail. You talk too loud and they send you off to see the warden. They're looking for us to make mistakes. The teachers, the principals, the cops, they got their eyes on us. If I told a cop I was going to college, he'd say, 'Who are you trying to kid? You're only a dumb punk.'

"That's the difference between the kids here and the kids in those fancy schools in the country. They don't have punks out there. All the punks go here. You never even hear the word *punk* out there. They got kids who accidentally misbehave, but no one over there is a punk. You got to be poor to be a punk. I'll bet even if I go to college, and then medical school after that, people will still call me a punk. 'Hey, Doctor Punk. Come over here and look at my kid's broken face.'

"What gets me is that they have it so good, there's no way one of their kids is going to end up here. And they fix it so none of us kids get to go there, unless we marry one of them. The way I see it, my education is only a ticket out of here. If it's really good, I'll take my whole family with me. If not, I'll go by myself. Maybe go find a rich girl. When I was a kid, I thought the most important thing was to stay around here and help my folks. But when I see what those fancy neighbors got, I tell myself, *I* come first. The first thing is to get *me* out of here. I won't end up anywhere if I stay here, but here. And here is nowhere. Here is nothing. It's just hanging around getting ready to get yourself a crumby job and a wife so you can have lots of children so they can go to the same crumby school you went to so they can hang around going nowhere. Then everybody will talk about me like I talk about my old man. Him, me, and my kids, all those years, and nobody's moved more than ten blocks in all that time."

3. The Scope of the Present

In part, a present orientation involves reflecting on the past and expecting the future. Partly too, it involves the experiencing of ongoing events. The meaning of a present orientation ranges from the instantaneous present, a perception we call scientific and depersonalized, through a slightly longer present in which personal action is partly defined by social aspects of time, like a day's work, and finally the longest duration of the present, a duration defined in part by sociological and cultural features, like the tenure of one's work. Thus, one could say that the present's duration is as long as a lifetime or as long as a culture survives. There is also a spiritual sense of the present to which Carl Jung (1953) alluded: "The man we call modern, the man who is aware of the immediate present, is by no means the average man. He is rather the man who stands upon a peak, or at the very edge of the world, the abyss of the future before him, above him the heavens, and below him the whole of mankind with a history that disappears in primeval mists" (p. 196).

In perceiving the present as instantaneous, we do not know whether a person is suggesting that experiences are felt or merely observed. That is, we do not know whether the person believes he or she is causing events to happen, or whether events happen that then affect the person in some way and in turn influence the person's perception of the present's duration.[6] We have spoken of the ways in which people believe they can shape the content of the future. Now we realize that in deciding how long the present lasts, people are shaping its content as well.

What we are suggesting here is that the perception of the present as instantaneous implies that the perceiver is really an *observer* of action. The perception of the instantaneous present makes us think of the person standing back from his or her own life and watching events pass by. In contrast, perceiving the present as *extended* implies that a person is the *agent* of action, the person making events happen. What this person observes is not the flow of events but the *effect* he or she has on the shaping of the present. This is the person Michael Dummett had in mind when he spoke of taking responsibility for one's life. It is in this sense of agency, this sense of taking responsibility, that one develops an understanding of personal autonomy and competence (see Erikson, 1968; Lewin, 1935; White, 1959).

The complexity of the perception of the present is still not fully appreciated, for we know that the sense of responsibility for one portion of the present may not be as great as the sense of responsibility for other portions of the present. Let us think of the concept of an *asymmetric* present. While one man may be preoccupied with the *past* portion of what he calls the present, another man may be preoccupied with what he calls the *future* portion of the present. Thus, people may be pre-occupied with what is just about to occur in the present or what has just occurred in the present. But notice that the word *just* refers to an amount of time for which the person presumably feels some sense of responsibility.

[6] This notion is based on Michael Dummett's distinction between the agent and the observer of action. See Dummett's "Bringing about the Past," in Gale (1967).

In perceiving the present as extended, moreover, people have a sense of where the past and the future connect with the present. But the boundaries between the present and the past on the one hand, and the present and future on the other, may be very unclear. It may be difficult for a person to say exactly where the past ends and the present begins. For some people, the temporal horizon is seen as a series of disconnected periods of time (see Murray, 1959). One thinks of these disconnected time periods when examining a photograph of oneself as a child. The photograph reminds us of a period of time called the past, which, while we know it is linked directly to the present, is nevertheless felt to be a piece of one's childhood, seemingly detached from anything we would now call the present.[7] As strange as it seems, we may feel that 15 years ago is farther away from us in time than a period 30 years hence, because of our present activity, the future does not seem that far from us. The extended present, therefore, represents the expanse of time that a person believes he or she has shaped or will shape. What one observes as a portion of past time, the content for which one feels responsible, and what one observes as a portion of future time, the content of which one believes one can shape, are brought together and the entirety is called the present.

An 18-year-old man saw himself when he was young. His family had just come back from a neighbor's house where they had gone swimming in a backyard pool. When they reached their own home, his mother had made him shower. When he was done, he discovered there was no towel. Believing himself to be alone on the second floor, he ran to his parent's bathroom to fetch a towel from the linen closet. Opening the bedroom door, he saw his mother standing naked. Startled, she reached for something with which to cover herself. The young man ran from the room unable to stop thinking how unattractive his mother looked. He could not forget the sight of her.

"Saw everything, man," he said to an invisible companion in a cowboy drawl. "Older women ought to be sent out to pasture with horses. They can make you sick just looking at them. All they got is a few years, man. When's this pill going to do something, for God's sake?"

---

[7] The example of the photograph of childhood illustrates the fact that the concrete nature of already-experienced time may seem less real to some people than the time they imagine through their expectations and anticipations. The example of the photograph is also reminiscent of Erik Erikson's work on identity (Erikson, 1968).

"Take it easy, man."

"Take it easy yourself. You don't *get* it, man. You don't get *nothing* about living and dying. You can't get on top of it. I can see it, man. I can see the bottom and the top of it, man. Way over the top, into the future, man, and as far back into the past as anyone would ever want to see. Don't tell *me* to take it easy, man. I can see it all, man, and you can't see a thing. All you see, man, is school and work, mommies and daddies. People having babies, man. You can't see it up top, man. You understand seeing over the top, man, into the future, man? Don't give me that high-school coaching stuff: take it easy. I *am* taking it easy, man. I'm flying, baby. You can't do that. You don't get a word I say, man. I'm up there, man, over the top, into the future. Way the hell over the top. I *will* take it easy. Real easy. Nothing hard now for me, man. I got it now. You don't think I do. I'll give it to you, man. You just behave yourself, keep quiet, stick close to me, man, I'll give it to you too, one of these days. You just behave, man. Keep your beady little eyes on the master. You want money, man? I see the top of your future like a little ant walking around down there. You scared, man? You ain't? Sure look like you're scared out of your mind to me. Get above those trees now, man. You ever see a man fly? Well, feast those little eyes of yours on this body right here. Go on, take a look at the big bird. Let's hear it, everybody, for the vertical man."

The young man ran his hands hurriedly over his chest and stomach and genitals. His body was covered with sweat, and he had become hoarse.

"Hey, God, man," he whispered, the vessels in his neck straining, "don't let me go crazy. You hear me, man? I'm not going crazy. You hear that? Answer me, for Christ's sake. I'm not going crazy, right? RIGHT? RIGHT? ANSWER ME. PLEASE ANSWER ME! TIME *IS* MOVING, AIN'T IT?"

4. Fantasies of Time and Death

The distinction between linear and spatial conceptions of time passage are mainly distinctions between the objective and subjective perceptions of time passage. We know that once experienced, a moment moves into the past, never to be experienced the same way again. Still, we feel that by recalling the past, we return it to the present. Thus, we *feel* we can reexperience the past.

Surely, there are instances when feelings take precedence over knowledge. One is when we fantasize about time, when our wishes, that is, govern our perceptions. The wish to recover the past and the wish to know the future are two such fantasies.

Recovering the past and knowing the future are fantasies almost all of us experience. They belong to the realm of imagination. We know these fantasies cannot become facts because we know objectively that time passes. Still, we entertain these fantasies, making ourselves happy or sad doing so.

Herbert Fingarette (1963) captured the vividness of time fantasy with his notion of "sense of presence":

> The sense of "presence," of nearness in subjective time is generated then, when any object or situation is cathected by the currently mobilized drives and when it plays a significant role in the dominant drive-fantasy complex. Such a "dynamic theme," or "unifying theme," once mobilized in waking life or in the dream, is like a magnet. . . . The current perceptions incorporated are then perceived as "real"; the memories, though perhaps locatable in long past (calendar) time, are, in subjective time, "as if it happened yesterday"; the hopes are vividly present: "I can see it already!" (p. 207).

Generalizing from Fingarette's statement, one senses a freedom in fantasy, a freedom that William Henry (1956) said, "derives from the less conscious and less structured aspects of the individual's personality. To these areas of personality the rules of logic and propriety do not apply" (p. 1). We are free to fantasize about the past and the future as we may fantasize about the historical past, the period of time preceding our birth, and about the historical future, the period of time following our death.

It is also true that fantasies about time may or may not build upon other people's reports and conjectures. Thus, we may distinguish between so-called *authentic fantasies,* fantasies developed in part from memory or expectation of personal experiences, and so-called *inauthentic fantasies,* fantasies evolving from secondary sources and experiences. These we have never known directly, nor will we ever know them (Merlan, 1947).[8]

[8] Medard Boss (1963) wrote about these notions of authentic and inauthentic fantasies in terms of the re- or preexperiencing of original time or life-space time, and the imagining of indirectly encountered history: "Original time is no external framework consisting of an endless sequence of 'nows,' on which man can eventually hang up and put into proper order his experiences. . . . Man's original temporality always refers to

Despite all the distinctions we may make between and among authentic and inauthentic fantasies and linear and spatial conceptions of time, the future still cannot be known or experienced until it arrives. The closest we can come to knowing the future is to imagine it. As Maurice Merleau-Ponty (1962) wrote:

> Reproduction presupposes re-cognition, and cannot be understood as such unless I have in the first place a sort of direct contact with the past in its own domain. Nor can one, a fortiori, construct the future out of contents of consciousness: no actual content can be taken, even equivocally, as evidence concerning the future, since the future has not even been in existence and cannot, like the past, set its mark upon us. (p. 413)

Still, imagining what the personal future holds can be distinguished from imagining what the historical future holds. If one knew the contents of the historical future, one would be able to evaluate such a thing as existence itself or the evolution of cultures. It would also be equivalent to gaining immortality. To know the historical future is to eliminate death.

For Husserl, death was not a moment in time and hence not a temporal event. Similarly, Martin Heidegger (1962) wrote, "The future is the ability to endure oneself in one's finiteness" (par. 65). Putting it as Hegel did, "Man pays the price of his individuality by being capable to contemplate death" (in Meerloo, 1966, p. 240).[9] The following passage was written by Cournot:

> Religious manifestations are the necessary consequences of man's predisposition to believe in the existence of an invisible, supernatural and miraculous world, a predisposition which it has been possible to consider sometimes as a reminiscence of an anterior state, sometimes as an intimation of a future destiny. (in Unamuno, 1954, p. 211)

> And it is this problem of human destiny, of eternal life, or of the human finality of the Universe or of God, that we have now reached. All the highways of religion lead up to this, for it is the very essence of all religion. (in Unamuno, 1954, p. 223)

---

his disclosing and taking care of something. Such original temporality is dated at all times by his meaningful interactions with, his relating to, that which he encounters. Every 'now' is primarily a 'now as the book is missing,' or a 'now when this or that has to be done'" (p. 46).

[9] According to Mircea Eliade (1961), the historical past and the future are said to be the time of the *sacred,* in contrast to the personal past and future, which are said to be the time of the *profane.*

A preoccupation with death has arisen in the minds of many young people. Death is talked about, witnessed, found to be repulsive, beautiful, scary, fulfilling. War fans the winds of death. Assassinations do, too. Wretched drug experiences and psychoses belong to the family of death, and so, too, ironically, does sleep. One cannot say that *only* the young contemplate or fear death. This is not true. Still, our culture has made such an outrageous commodity of youth that childhood and adolescence become especially difficult times to relinquish. And because they are, death suddenly becomes a vital part of growing older.

"You realize I'm only sixteen and I think about rotting away, disintegrating every day of my life?" He is a student from a working-class high school. "I'll be driving in someone's car or looking out the window coming home on the trolley. Just looking out the window, and I'll see some old man hobbling along on the street, leaning on his cane or being helped along by some other old man, the two of them not even sure just who's holding who up, and both of them like a couple of drunks about to fall flat down on the street. Right out in the open. And I'll be looking at them as though the trolley wasn't even moving at all. Like everything was standing still. Then, like, I'll make myself think I am one of those old men, dying right there. Down on the street, splitting my head open and lying there, with the cops coming and dragging me off in that special wagon they got for corpses. I can't get myself to stop thinking it. Only way I can stop is to look down at my hands and just make sure they're still young looking. Not wrinkled or anything, or all ripped up."

Some think of youth as the final period of dependency and protectedness. Others describe it as the storm before the lull. Whatever its psychological components, youth clearly is affected by what we often smirkingly call "adult maturity" or the "established way." Many young people ask themselves about "making it" in one manner or another. Some have sought new modes of behavior, modes that bypass traditional achievement-ventures and the anguish associated with careers. But the notion of survival pervades these new modes.

No doubt the young care about the destiny of their culture, of their families, and of their private ways. No doubt they fear that life can end at any time merely because those in power might find it necessary or inevitable that some of the young should die. And so it appears that some young persons—how does one say this?—identify with death.

They seem to be inspecting the vulnerabilities of their society, their bodies, intellects, and emotions. Some even feel a sense of responsibility for the fact that our culture may not endure, that time may stop.

"I can see myself as life or death, if you know what I mean," a young woman from Kansas told me. "I carry around the feeling that if somehow, in some magical way, I could find a reason to be living, then I could keep myself alive, kicking, breathing, sexing. And then, in that same stroke of magic, I'd be able to save us all. The children, the very old, too. I don't mean anything as silly as some Christ image. It's just that each person must take his own life in his own hands, hold it there, inspecting it, you know, like a new piece of jewelry, and then make sure nothing happens to it. I guess I feel that if I could do that, then every one of us, no matter where they are, or how sick they are, or how discouraged they are, just might, you know, survive."

From a 19-year-old man: "I hate to admit certain feelings, but I wish for people to die a lot. Almost everyone I have ever known, even the ones I like most, I spend time thinking about them dying. Then I think about my own death, or being old and getting close to death. I often imagine myself very old making a list of all my regrets, all the things I wanted to do when I was young. I don't really know what I want to do. I want to know. I know deep inside I can't make significant decisions. I'll go to my grave not knowing what I wanted to do or even what I regret not doing. While I'm alive, I just want to stay away from death. When I'm dying, I'll want to die, but I guess I won't know how. I can't explain it any better. I just can't live in the world the way I am. It's not that I was born too soon or too late. It wouldn't have mattered when I was born. As long as I am me, I can't fit in. Death comes never too soon!"

5. Life's Passage and the Varied Perceptions of Time

We have now discussed a great many perceptions of time. Some of these perceptions may seem relevant to the way our lives are led, while other perceptions may seem relevant to little at all in our lives. Accordingly, we must ask, does one's perception of time help a person to understand and appreciate the nature of living and dying?

One of the major features of the ego, Lucille Dooley (1941) wrote, is to ward off the anxiety that comes from recognizing one's impermanence. One way to ward off the anxiety is to view the passage of time in such strict chronological terms that one begins to forget that his or her own life is running out with each passing moment. Another way to ward off the anxiety is for the ego to fragment the passage of time into minute units, each presumably capable of sustaining one's belief that every individual unit can last forever. As some people would say, I live only for today, I don't think about tomorrow or anything's coming to an end. These people may achieve the feeling of *permanence* by carving out a series of present-time zones, each one so long it seems that they could not possibly live long enough to reach the end of any or all of them. Finally, people may ward off the anxiety about the impermanence of life by placing personal experiences in the contexts of historical ceremonies or even the occult. In all these ways, we become involved with aspects of time that help us to understand the meaning of a single life and ease the fear of a single life's lasting for so short a time (see Becker, 1974).

In the end, these aspects of time perception are of the utmost importance to people as they lead their lives day by day. All human beings must work out their own conceptions of time flow and their own perceptions of the temporal horizon. They must deal with their own pasts, their own presents, and the past that belongs to history—the time that existed before their own births. They must deal, moreover, with their personal future with its unknowable content, just as they must deal with the historical future, a period of time that they will never experience. And finally, all people—influenced by social, cultural, religious, and educational systems, by their own personalities, and by the rules governing their social roles—must attempt to understand and make peace with the reality of their birth and their death.

These are quintessential life concerns, concerns that must be at the forefront of any examination of time perceptions. But a study of people's perceptions of time is made even more complicated by the fact that these perceptions constantly change because people are living in time. With each second, our perceptions of the past, the present, and the future, if only in chronological terms, must be altered. At the conclusion of his prologue to *Identity, Youth and Crisis* (1968), Erik Erikson wrote that the study of human lives must at some level be a study of biographies in history; people do not hold still long enough for anyone to describe them. Similarly, a study of perceptions *of* time is a study of the movement of human beings *in* time. People are always in motion and that very motion is part of the horizon of time people experience.

All research on human beings must take into account this movement of life and the constant redefinitions of oneself that go with growth and maturation. The experiment in which the investigator examines before and after conditions is an example of research in which changes and development over time are being explicitly studied. Whether it is mentioned or not, time is a major variable in this type of study. But when research is aimed directly at perceptions of time, then we have complicated our task, for we have focused on what for many researchers is only the background of their work. While some researchers may want to know how people have changed from one period of time to another period of time, we may want to know how people perceive this period of time, in addition to how they perceive changes in themselves *over* this period of time. People's perceptions of the passage of time become for us a major determinant in shaping their perceptions of how they may have changed over time.

"My talk about dying," a young woman of 17 said, "is a lie. I don't want to die. I want to live, no more now than before, no more later than now. I hide in death because it seems safe, but if it were truly safe I could imagine myself being no more, and I cannot. I can only imagine lying somewhere looking up, knowing that it is quiet. But I always see the sky and the clouds moving, so I know that I am not dead, not even in imagination.

"People who work at preparing themselves for suicide imagine being no more. If they cannot, then they can think of being no more and not let if frighten them. If they cannot do that, they can think about death and of someone they love in the same breath and not worry that they might die and that the person they love might live. When jealousy has vanished from a person's mind, when they no longer resent others' living when they are dead, then they can accept dying.

"Living and dying, people think, are done by oneself, only by oneself. But it isn't so. We do nothing alone. My mother and father are part of my studies of life and dying. I might die tonight, this very minute, hundreds of miles from them, but they are part of my death. They are part of my living, part of my dying. They will be a part of my dying if I die years after they are dead. My parents alone have created that in me. I am a child unable to move beyond the youngest age I shared with them. I am my parents' child. It is the dilemma I have inherited from an endless line of women. I am a child, created by them, kept that way by them, no matter the distance between us.

"I fear aloneness more than I fear dying. One separates from people, from parents, and one survives. I draw a knife over my skin and see blood pouring out of the separation I have made in my own skin, and I survive. Eventually, one even separates one's past life from one's present life, and survives. The body and the spirit separate, they tell me, and one or both of them survive. So what is to fear? Death? Aloneness? One's imagination? *I* create the line that separates me from them, that which I conceive from that which the world thrusts upon me, the people I come from, from the people I shall give birth to."

References

Becker, E. *The denial of death.* New York: Free Press, 1974.

Bergson, H. *Matter and memory.* New York: Humanities Press, 1964.

Boss, M. *Psychoanalysis and daseinanalysis.* L. B. Lefebre (Trans.). New York: Basic Books, 1963.

Cassirer, E. *The philosophy of symbolic forms. Vol. 3. The phenomenology of knowledge.* New Haven, Conn.: Yale University Press, 1957.

Dewey, J. *Human nature and conduct.* New York: Modern Library, 1957.

Dewey, J. Time and individuality. *In* H. Shapely (Ed.), *Time and its mysteries.* New York: Collier, 1962.

Dooley, L. The concept of time in defense of ego integrity. *Psychiatry,* 1941, *4,* 13–23.

Eliade, M. *The sacred and the profane.* New York: Harper and Row, 1961.

Erikson, E. *Identity, youth, and crisis.* New York: Norton, 1968.

Fingarette, H. *The self in transformation.* New York: Basic Books, 1963.

Gale, R. M. *The philosophy of time.* Garden City, N.Y.: Doubleday, 1967.

Gioscia, V. Classification of family types. Unpublished manuscript, Queens College, City University of New York, 1966.

Heidegger, M. *Being and time.* New York: Harper and Row, 1962.

Henry, W. *The analysis of fantasy.* New York: Wiley, 1956.

Jung, C. G. *Modern man in search of a soul.* New York: Harcourt, Brace, 1953.

Kant, I. *Critique of pure reason.* New York: E. P. Dutton, 1934.

Kummel, F. Time as succession and the problem of duration. *In* J. T. Fraser (Ed.), *The voices of time.* New York: Braziller, 1966.

Lewin, K. *A dynamic theory of personality.* New York: McGraw-Hill, 1935.

Lewin, K. *Resolving social conflicts.* New York: Harper and Bros., 1948.

Meerloo, J. A. M. The time sense in psychiatry. *In* J. T. Fraser (Ed.), *The voices of time.* New York: Braziller, 1966.

Merlan, P. Time consciousness in Husserl and Heidegger. *Philosophy and Phenomenological Research*, 1947, *8* (1), 23–54.

Merleau-Ponty, M. *Phenomenology of perception*. New York: Humanities Press, 1962.

Murray, H. A. Preparations for the scaffold of a comprehensive system. *In* S. Koch (Ed.), *Psychology: A study of science, Vol. 2. Formulation of the person and the social context*. New York: McGraw-Hill, 1959.

Schutz, A. *Collected papers. Vol. 1. The problem of social reality*. The Hague: Martinus Nijhoff, 1962.

Unamuno, M. de *The tragic sense of life*. J. M. C. Fitch (Trans.). New York: Dover 1954.

Van der Leeuw, G. Primordial time and final time. *In* J. Campbell (Ed.), *Man and time*. New York: Pantheon Books, 1957.

White, R. W. Motivation reconsidered: The problem of competence. *Psychological Review*, 1959, *66*, 297–333.

Whitrow, G. J. *The natural philosophy of time*. New York: Harper and Row, 1961.

6

Editors' Introduction

As we look at the elderly person, the traditional categories of "past," "present," and "future" might seem to have a certain established meaning. Older people, having lived out most of their lives, are apparently "running out of time." Therefore, we might be led to believe that they are predominantly past-oriented and that their concepts and projections of plans into the future may be limited.

Robert Kastenbaum, an authority on the psychology of aging and dying, dispells these simple notions and stereotypes and reveals that although we may tend to think of the future as "ahead" for the young person and the past as "behind" for the older person, in their individual modes of existence this may not necessarily be the case. Kastenbaum shows us that there are many ways in which older people use concepts of their past and future lives, and he offers a delineation of the varied modes of experiencing the past and future.

Kastenbaum shows that, in fact, some elderly people seem to be far more future-oriented than many younger people. For some elderly people, past orientations provide joyful activities that transcend the dull roles often accorded to old age. The old person, like the young person, can employ different and highly individual time perspectives as a reflection of his basic life style, including his philosophies and cognitions of purpose, continuity, and utility.

6

Memories of Tomorrow: On the Interpenetrations of Time in Later Life

Robert Kastenbaum

Why would you and I be here today unless each of us had some private fascination with time? And how would we have anything substantial to share with each other unless, over the years, some of this private fascination had been transformed into controlled observations and all the rigorous maneuvers of scientific theory and research? Yet I continue to wonder how well our studies represent the experiential world of psychological time. Consider, for example, the meanings and uses of time for those who have themselves survived the passing of many hours, years, and decades. It is doubtful that either popular stereotypes or the available research do justice to the varieties of time experience at their command.

Today's choice gathering of psychochronophiliacs encourages me to share some puzzlements, observations, and ideas in the general area of time and old age. Not much attention is given here to material that is already in print, whether yours, mine, or somebody else's. We can all read the literature and come to our own conclusions. Instead, I invite you to step in and out of the minds of a number of people whose everyday experiences with time have filled my own mind with more questions than I can answer. Perhaps you can comprehend these phenomena. If so, marvelous! If not, then at least I will have some good company in my present confusional state.

Robert Kastenbaum • Department of Psychology, University of Massachusetts—Boston, Boston, Massachusetts.

1. Images and Vignettes

Let us begin with a few brief images and vignettes, all of them well within the ordinary range of everyday experience.

1.1. Three Fishermen

First, I will tell you something about three fishermen, using this term a little loosely. One of these people is in a state of rapture or exultation; it could not be described with a lesser term. He has been wandering around the Fish Pier in Boston. Everything associated with fishing has excited him—the small crafts bobbing at anchor, the lobster traps, the paraphernalia of deep-sea fishing, all the sights, the sounds, the smells, the people, and, of course, the fish themselves. He has been especially taken by a marvelous room he discovered in which men are doing all kinds of interesting things to all kinds of interesting fish. David is usually a person of action, living each day fully and not much given to future projections. But this morning's experience has awakened something within him. "I would *love* to work here!" he confides, and he repeats this several times. How old would he have to be to get a job on the pier? Who should he apply to? Could he work on the pier, doing something with the fish, and still go to college? Shining in the eyes of this 11-year-old is a pure and transfiguring light, an expression of sacred purpose and rapture almost unbearably keen to behold.

The second fisherman is flat on his back in a hospital bed. He has been old and sick for a long time; certainly for as long as I have known him. Now he looks not only old and sick, but also unhealthy. His skin is pale; his eyes appear to be floating at the surface of his face; his breath comes in gasps and gulps. In contrast with his physical appearance, however, the mind is focused and in control of itself. We have spoken together about sickness and pain, life and death, and we will speak about it again. But just for the moment, he has something more important on his mind. This is the day the fishing season begins, isn't it? He is right about that; how could he have known? He speaks about local fishing places and seems on the verge of expressing either the desire or the intention to take his pole and reel to a favorite spot. Instead, however, his thoughts turn pastward. He recalls a short vignette that proves a long time in the telling, for he is now more relaxed, his breathing partially smoothed out, and words and phrases come in widely spaced groupings. The vignette centers around a day when his father took him fishing . . . how he can still remember all the details and, especially, how proud it made him because it meant that he was old enough to be a companion for his father, who usually preferred to do his fishing alone. After that, he went fishing many times with his father, and those were good days. He is a serene person during this shared recall. The impression I have is that he is at the same time both "living in the past" and completely aware of his current situation, including his life-threatening illness and our immediate interaction. His pale body, restricted to a hospital bed, somehow appears irrelevant to the man and his experiences.

The third fisherman is on his feet and busy. He is operating a small, assembly-belt kind of apparatus and has a mountain of fresh, slippery fish to process. Hundreds of fish eyes are upon him, but he does not return their glance. This man is obviously a pro at his job. You just know that every fish will be packed firmly into its appropriate box right on schedule, including the two that have skidded off into the corner and seem to have eluded his notice. There is nothing about his attitude you would call sullen or lethargic, and we do not know what he actually has on his mind. But his actions are entirely routine and matter-of-fact. Only for a moment does an expression fleet across his face—as he becomes increasingly aware of David's admiration, flushes slightly with pleasure and embarrassment, and gives the boy a friendly nod.

The fourth fisherman—this is a bonus; better than promising three and delivering only two!—has a paint brush in hand. He is painting a fish. I mean, he is literally painting a fish, not a picture of a fish. Both he and the fish are nude. Both are being swarmed by black, biting flies. The painter works intently, delicately outlining the scales of the dead fish, despite his sweating profusely in the summer heat and undergoing the seige of the flies. He does not allow himself the luxury of shooing away the flies or wiping the sweat from his face. Finally, it is the smell of the rotting fish that impells him to terminate this activity. But as he steps back to admire his achievement, the painter thrills to a discovery. It is not only the fish who glitters and sparkles in the sunlight: "Oh! I was covered with shimmering scales! But at the same moment I realized where they came from: it was only the dried splashes of my crystallized fixative. . . . I stayed by myself, dreaming till dusk. O Salvador, your metamorphosis into fish, symbol of Christianity, only came about because of the torment of the flies, what a typically Dalinian crazy way

to identify yourself with your Christ while you are painting him!"
(Dali, 1965, pp. 32–33).

Tell me: why is it that we find rapture, serenity, and mystical discovery in the experiences of a boy, a seriously ill old man, and a highly
individualistic artist—but no such state in the actual ongoing behavior
of the one person who is at the moment an active fisherman? Does this
tell us something about the relationship between prospection, retrospection, and the present moment that we perhaps understand "intuitively"
but have not incorporated into our theory and research? David's present
rapture has been occasioned by his future projection of life on the Fish
Pier. Yet there is no rapture, serenity, or mystical participation to be
seen in the eyes of those who actually labor on the pier today. Only the
fish look as though they might have a special experiential state, and
they are not talking.

Anticipation and retrospection are often regarded as mental
strategies, if not as defense mechanisms, while "real reality" happens
right now. We are familiar with the position that only the now exists.
On this view, it is acceptable to study past- and future-oriented modes of
thought for what can be learned about individual or social dynamics.
However, the past and the future should only be nice places to visit; we
are expected to keep our ontological noses to the grindstone of the
moment, even if we are never told precisely why. As a corollary, we are
expected to remain finely tuned to the *passage* of time. Follow the
bouncing ball! But my satisfaction with this view continues to be
undermined by experiences with the very young and the very old, as
well as with some of us in-betweeners. A few more examples either will
be instructive or will, at least, add diversity to the gathering confusion.

1.2. A Bouquet of Futures

Let us now make a bouquet of futures by taking just one or two
offerings from a variety of people. In each instance, I have tried to select
a central or thematic representation of the individual's overall time
perspective, but this must be recognized as a simplification procedure.

Mr. R. is in his 80s, resident of a medically-oriented geriatric
facility where he has lived for several years and which he expects to be
his final abode. This is a man who keeps to himself but also keeps a
sharp eye on what is going on around him. His most characteristic
expression is along the spectrum of cantankerous, stubborn, displeased,

disdainful. And yet, he has become a respected and fairly popular figure within the institutional community, perhaps because of an aura of integrity and honesty, as well as an occasional humorous observation. What of his future? Mr. R. does not have one of those. Hours of personal contact confirm the impression that he considers himself a person with nothing ahead of him that could be called a future. He does not waste his energy or kid himself by thinking ahead.

In this sense, then, Mr. R. stands as a typical representative of the old person some of us have studied and written about. Give him a formal time perspective procedure, as I did, along the way, and sure enough, he will prove to have a very limited orientation toward the future. But this does not necessarily signify anything about his *age*. Had we inquired about Mr. R.'s time perspective three or four decades earlier, it is likely that the same orientation would have been discovered. Within his own frame of reference, Mr. R. "had prospects" only once in his life. After a struggling childhood and adolescence, he had one big shot at success and financial independence. He put everything he had into it, but there was a terrible frost that year, and his potatoes froze and rotted in the ground. Since that time, he has had no future. There has been no projection of plans, no sense of moving onward, just a continuing-to-exist. The future is something he went past a long time ago, and we needn't blame old age for that.

Ms. T. lives a tidy life and keeps a tidy apartment. She continues to work, although she is approaching her mid-70s. Apparently, she remains a valuable employee in a family business, the person who has been around the longest and can be counted upon to know and do the right thing. This is a woman I have not actually met myself, but through the observations of a student in my aging class. Each student is expected to become companion to two elderly people as part of the course requirements. The student who became friendly with Ms. T. gradually became frustrated with this woman's style of life. She complained that Ms. T. did not really know how to live. There she was, still working hard and still scrimping and saving money. Yet she was in good shape financially. She didn't have to work, certainly not as hard as she did. What in the world was she waiting for? Why didn't she enjoy herself while she still had time and health? The student wanted me to suggest a technique that she could "use" on Ms. T. so she would start getting more out of her life.

Instead, I proposed that the student recognize and set aside her

own values for the moment and observe Ms. T.'s life style more carefully, offering a few suggestions for doing so. A few weeks later, the
student had a different view of the situation and, to some extent, of
herself. She now saw Ms. T. as a person who is essentially continuing
to do what she is "good at." It is not just that Ms. T. is a good and
valued worker. But it is also that she puts her heart and mind into
scrimping and saving. Theoretically, the savings are intended for use
someday. But "someday" no longer means what it might have many
years ago. The future as a consummatory state, a time–place where the
virtuous life is rewarded, has, in effect, withered on the vine. What Ms.
T. has in its place is the firmly entrenched, well-practiced knack of
preparing herself for the future. The routine of daily life only seems to
borrow its significance from the promised vista of rest, pleasure, and
fulfillment. In actuality, however, it is life-as-preparation-for-future, not
future-as-reward-for-virtuous-life, that has become the organizing principle for Ms. T.

Mr. L. has something important in common with Ms. T. He has
also spent most of his years on this planet in hard work and self-denial.
Nothing came easy for him, but he kept plugging away. The details of
his life differed considerably. He did marry and help to raise a family.
Mr. L. characteristically worked at two jobs to support not only his
immediate family but also various relatives who came to depend upon
him. He took his obligations toward both the past and the future
seriously, spending most of his time and money on making things comfortable for the older generation and easier for the younger generation.
For himself, "Why, I expected I would have my day, someday!"

His day is now one of crippling impairment and semiabandon-
ment. He resides in a geriatric institution, although the staff believes he
could live at home with reasonable assistance on the part of his family.
Mrs. L. has taken control of the family constellation and seems to have
no place left in her life for this worn-out, no longer productive person.
The children have their own lives and put in only token appearances. It
is only with great reluctance that Mr. L. will now share the sustaining
ambitions and fantasies that he entertained throughout the years. He
also attempts to conceal a fierce bitterness about his fate beneath a tran-
quil, philosophical exterior. But the anger does flame out on occasion,
followed usually by a surprisingly youthful expression on his face. He
manages to look somehow like a young boy who has done everything he
was told to do but who has been deprived of his hoped-for prize. When

he describes himself as a son, a husband, a father, and a working man, uppermost is the theme of being "good" and "responsible." Now he exists in a stunned and bitter condition. Unlike Ms. T., he has not relinquished the image of a future that should make all the effort, all the self-denial worthwhile. Mr. L. feels that he is entitled to love, admiration, and respect, earned by so much dedication and hard work (although he admits that he probably has been something of a tightwad and a grouch in his family's eyes, even though it was only for their own good). He also feels that God should be giving him time that he can do with as *he* wants to, not this restricted life tied down by an impaired body in an institutional setting.

In a rare moment of expansiveness, he will share his fantasy of a time still to come in which everybody will realize what a wonderful person he is and has been. Curiously, this future will "make everything all right" by focusing much upon the past. Tomorrow, in other words, he will be able to share his past with family and friends who will then cherish all he has done to make their good lives possible. Although Mr. L. probably has been as diligent and scrimpy as Ms. T., he still wants his future and seems ready to cast aside the so-called Protestant ethic in favor of self-indulgence if only given the opportunity.

While Ms. T. comes across as a "completed person" in her own way, there is something inconclusive about Mr. L. He still "has prospects," or, at least, so he permits himself to believe at times. He still can go through the motions of the carrying-on-as-usual, stalwart, self-denying individual—and yet again, he is a bitter man who might eventually give way to his gathering resentment and frustration by overturning his faith in God as well as in his fellowman. There are multiple lines of futurity, then, still functional within Mr. L., and each has different implications for the meaning of his past life as well.

Anxiety rather than bitterness is a keynote for Mr. G. Strictly speaking, he does not qualify for our attention here because he is still in his early 50s. But Mr. G. has his mind very much on his old age, and beyond. He is a well-to-do person who has developed a base of social and informal political power. People tend to do what Mr. G. wants them to do. But his own children, a son and a daughter, have failed to meet one of his major expectations. Mr. G. feels that he cannot rest easy in his mind until there are grandchildren on the scene. As a matter of fact, he fully intends to remain alive until great-grandchildren also check in. He has been pressuring his children for years. Most of the pressure is upon the son. When Jimmy was coming into manhood, his romantic adventures delighted his father (a chip off the old block, you know). But now it is high time for Jimmy to settle down and raise a family. Although Jimmy has proved obedient to his father's wishes in general, there are still no signs of genetic succession upon the horizon.

Mr. G. will tell you openly about his desire to bestow both worldly goods and the good family name upon the fortunate future generations—if only they would come forth! He will also add that he is a man at peace with life and death. He has been living a full life; he has no regrets; he could die tomorrow or even today with a smile on his face. Yet there is an edgy, driven quality to these remarks. Mr. G. represents a tradition that might be dwindling away rapidly in our times, the tradition of a "family line" in which the name and therefore the "family soul" is preserved through a kind of sociogenetic immortality. In this tradition, the individual's death fear, especially if the individual in question is the patriarch, is best alleviated by his passing the torch on to the next generation and beyond (Kastenbaum, 1974). The future that really concerns Mr. G. is not his own old age as such but the image of time continuing far beyond the visible horizons, with his identity, values, and name carried forward. I believe he would prefer to die today with a guarantee that his name and values will be carried forward indefinitely in new generations than to live another three decades or more in prosperity and health.

Even more briefly, let us drop into the minds of three other people. Miss B. and Mrs. Q. are both active and vigorous women, especially when their ages are taken into consideration. Miss B. is a tiny, dynamic person who seems to take each day into her own hands. The only time I have seen her even momentarily derailed by inner, psychological concerns was near the end of her 89th year. She was genuinely afraid of turning 90: "That's old, don't you know!" A few weeks later however, she had discovered that 90 didn't feel much different from 89 and resumed her accustomed life style. Miss B. seems to have been a self-confident, independent, gregarious person all her life. Age has made little difference in this regard. She does not think much about the future, but then, she never did! Although a rewarding person to know, Miss B. proves frustrating to anybody who expects to find an elderly edifice haunted by ghosts of the past or attuned to the mysteries of the future. She is just a pragmatic and vital woman. She would not even think of saying that she lives one day at a time, because the idea that there is some alternative game plan never seems to have entered her mind.

Mrs. Q., in her early 80s, is also a bustling person who is close to the center in the sphere of local activities. She is on the phone and on the go much of the time. What makes her orientation appreciably different from Miss B.'s is something that must be interpreted in terms of recent events in her life. Mr. Q. suffered an industrial accident and underwent amputation of one foot many years ago, while their children were still young. His general health declined after a while, and for about two decades he was a semi-invalid. Mrs. Q. served as more of a nurse than a wife. Both eventually became patients in the same geriatric hospital. They spent much time together, still organizing their lives around each other. When Mr. Q. died, some of us were concerned about Mrs. Q.'s will to live and her own survival. We needn't have been! By observation and by her own comments, she is flourishing more than she had been for years. Time is finally her own again—she is no more the mother or the nurse. She is "making up for lost time" by throwing herself into activities and relationships. She sees herself as being freed from the burden of the past and now able truly to live in the present. The future is somehow not very important just now.

The first time I saw Mr. J. was during my class on the psychology of aging. The discussion that day was shifting from an examination of the creative process in later life to intimacy and sexuality. One of the most intent listeners was this elderly man, who had suddenly materialized in the classroom. At the end of the class, he came up and introduced himself, in the company of an attractive coed who had suggested to him that he might enjoy the class and that I wouldn't mind having a guest. It turned out that Mr. J. was an octogenarian who looked at least 20 years younger than his actual age. He was also a poet. The highlight of the next class session—and probably of the semester—was Mr. J.'s reading of his own poetry. The reception was enthusiastic and touching. After a pause, he remarked shyly that he also had some erotic poetry and, with due urging, reached into his other pocket and produced a rollicking narrative on themes of love, sex, and nature.

This man is now well known in the community as a person and as a popular poet. He is having the time of his life. But he is actually working on his *second* life. His first life, as a political activist and commercial artist, ran out of future abruptly about 30 years ago. A temporarily incapacitating illness and other setbacks conveyed the message that his life, at least his productive life, was over. To keep up his own spirits and dispel boredom, he tried his hand at poetry for the first time. This proved a fascinating pursuit for Mr. J., and he, in effect, scripted a new life for himself after his original future expired.

What is it we can learn about old age and futurity from these people? Do we learn that old age "causes" one particular view of the future? Not likely. We see instead a variety of orientations, each of which must be interpreted both in terms of the individual's previous life experiences and the current situation. It would be a hollow triumph, if any, to conduct research so restrictive that the individual differences among these people disappeared, to be replaced by an essentially fictional set of conclusions about old people "in general." If there *are* generalizations to be made about the old person's relationship to time, these must be won by the most detailed, patient, and encompassing observations, rather than quick hops from stereotype to questionnaire and back to stereotype. It is possible, for example, that several of the people described here might give similar responses to time-perspective questions or tasks. And these might be indicative of a limited extension of thought into the future. Yet the context and meanings of these responses would differ appreciably—for Mr. R., who lost his future while he was still a relatively young adult; for Miss B., who has never troubled herself to think far ahead; for Mrs. Q., who has finally grasped the opportunity to live in the present; and so on.

For another example, consider the unknowns that are glossed over when we are quick to apply standard concepts and terminology. Both Ms. T. and Mr. L. are flesh-and-blood embodiments of the famous middle-class orientation toward work, virtue, and salvation. But what is to be made of the fact that Ms. T. is still *enjoying* her "delay of gratification" and perhaps will do so to the very end of her life, while Mr. L. feels embittered and cheated? Are the differences to be found only in the particular events that have taken place in their lives recently, or have these people made essentially different uses of the delay-of-gratification model in the organization of their thoughts, feelings, and activities throughout the years?

Indeed, I wonder if *any* questions about old age and futurity can be answered satisfactorily if our attention is limited either to old age or to futurity. Take Mr. G., for example. What is in this man's mind and heart at age 50 seems crucial to the meanings he will find in time at age

80. But Mr. G.'s time framework already has influences beyond his own person. In studying his relationship to time, we are also learning something significant about the forces that will be operating on Jimmy's relationship to time as he grows old himself—and perhaps it is not too farfetched to say that the grandchild who has yet to be conceived has a ready-made "time-force field" waiting to exert influence.

Could it be that time perspectives must be regarded as *inter*personal as well as intrapersonal structures? Is it unusual for several people to be involved in the same time-perspective network? Or is this actually a common phenomenon that we tend to neglect because of our interest in the individual?

I am wondering, then, how well prepared any of us are to do justice to both the interpersonal and the intrapersonal complexities of time perspective. The typical study seems to peel off several dimensions of the individual's future outlook and falls short of telling us either the role of futurity in the person's total configuration of time relationships or the ways in which his perspective interpenetrates with those of others. I am not criticizing this style of research as such but simply expressing dissatisfaction with it as the only or the predominant approach. Contacts with old men and women suggest that similar views of the past are not necessarily linked with similar projections into the future, for example. One person will summarize a long recounting of his hard and disappointing life by saying that there can only be rough times ahead. Life has been difficult so far, *therefore* that's the way life will continue. But another person will conclude an equally depressed view of the past on a much brighter note: things have been so bad that they will *have* to improve in the future! By knowing only a person's view either of past or of future, I do not have sufficient information to understand his relationship to time. Yet should I be fortunate enough to learn in detail of his more encompassing relationship to time, do I have the ability to discover how it all fits together—if it does, in fact, fit together—and to derive from this knowledge something that is a shareable contribution to science?

1.3. The Last Leaf and the New Blossom

The configurations of time perspective both within and between people should be illustrated more clearly. Let us turn again to studies from life. Here are excerpts from the time relationships communicated by two people.

The old woman has lived alone for many years; now she is in a congregate-care facility where theoretically there is much opportunity for companionship. Psychologically, however, she continues to live alone, as do many others in this facility. Perhaps we should specify that she lives alone with her thoughts—a peculiar kind of solitary confinement, if, indeed, it is solitary. Looking at this woman as the outsiders we are, we may not notice anything that sets her apart from others living in the same sociophysical environment. But she does have a distinction. She is a "last leaf on the tree." There is no futurity for her in the form of children and grandchildren. There are no survivors of the core family and the relationships from which her life took root. The present situation, then, in reality holds neither continued affiliations with the past nor the seeds of future continuation.

All the Last Leaf has left to her is the past that *is* past. And she is all that the past still has to rely upon. Generations remain alive only in the flickering memory of a person whose own days are drawing to a close. Her mind is the final common pathway, the last preserve of all that has gone on before in one branch of human existence.

Many such people are among us today, often not recognized as such. They seldom tell us; we seldom ask. These people carry a tremendous burden. I wonder how they do it. Imagine yourself to be the final survivor of the entire human race. When you go, so does everybody else. Dispassionate analyses of memory functioning would not catch hold of the significance of the images held in mind. You might well feel like a person who is bearing a burden incredibly heavy, yet incredibly rich. Your orientation toward death could not be understood simply in terms of individual feelings, which is our typical approach today. The impending death of the past through the demise of its host would have to be appreciated.

Examine the time perspectives held by the Last Leaf. Our traditional concepts and categories fall short of encompassing them. We find, for example, multiple representations of the self at various stages of development. Each of these self-representations tend to be linked with different constellations of people and situations. This means that "past," "present," and "future" are defined anew by each context that can be differentiated. Any given scene remembered includes not only itself but also its implicit past and future. Some past scenes are built largely upon future expectations that were dominant at the time.

We observe the host herself paying visits to these scenes remembered, stepping in and out of roles and, therefore, out of one time perspective and into another. Now she is the young child spoiled silly by grandparents, now the adolescent breaking away from home, now the grown-up at a turning point in life. People long deceased still look forward to futures that never were to be or have since gone by. Children still anticipate the new kitten in one scene, although it is buried as an old cat in another. Levels of reality vary within the scenes remembered: a probable versus an improbable future, both envisioned at the same time; a dubious story about what happened "in the old country" that took on its own kind of validity, and the true story that could only supplement but not supplant the previous account of the past . . . the deeper one proceeds, the richer and more differentiated the network of time frameworks. And yet, *all* these people, events, and relationships would be considered by outsiders as within the province of the past, as exhibits in a memory museum. The curator may herself be among the exhibits, but she also moves among us, just as subject to the ongoing flow of time as you and I.

This Last Leaf, Host, Curator, or whatever we choose to call her is of the opinion that she herself has no future. Yet it is her mind and her mind only that keeps so many past futures alive. Although in a certain sense, she does "live in the past," the past also lives in her, and the time dynamics may be as vital and complex as any that we experience within the consensual *now*. The past generations still retain their internal time relationships both within and between scenes remembered.

Perhaps it is important to distinguish futurity in the usual sense of the term from the memory of a future. We do seem to be referring to different phenomena. But the similarities as well as the differences are worth considering. Note, for example, that the *tension* of future orientation can remain sharp and experiential in memory. In other words, the "feel" of futurity can itself be represented in memory, that sense of distance, contingency, and movement that separates where we are now from where we might be next. Note also that some of our orientations toward the "real" future lack the qualities theoretically associated with futurity. If I am convinced that tomorrow will be just like today, then the future does not exist for me as time that is truly fresh, variable, new.

It is just like "any old time" and might as well be today or yesterday. Tomorrow is simply today happening again: this contrasts with the old woman's enlivened recollection of a moment in the past that still hums with stirrings of the future.

In practice, we may not only stereotype the future as being known in advance but may find discovery and adventure in the realm of the past. Whether it is an archaeologist literally digging into the past or two old-timers comparing personal memories and assembling a new history together, we can find many instances in which the past invites mental adventuring. The fact that one person is "future-oriented" while another is drawn to the past, then, does not necessarily indicate an open, questing mind on the part of the former or a closed mind on the part of the latter. This observation has probably occurred to most people, yet we tend to lose sight of it in a society that acts as though it places much more value upon future than past orientations. For example, we are apt to praise the youth who is thinking ahead and organizing his or her present activities around future prospects—but to put down the old person whose thoughts center around the past, even though his past represents a future actually achieved, in contrast to the youth's fantasies and ambitions that may never be translated into reality. The "last leaf" can recall generations of future strivings, some of which actually became the foundations for the realities of today. It is odd that we value the fantasies projected ahead and not the realities they eventually become, preserved in memory.

A person who qualifies as the Last Leaf by definition stands at the end of a long time sequence. Let us make a brief comparison with another person, one who might be called the "New Blossom." This could be an infant, but we will have more to work on if we move to a stage of development that has a clearer voice to speak for its experiences. Cynthia, a 7-*and-a-half*-year-old, was with her brother David on the pier when his own futurity became attuned to the fish and the sea. She enjoyed her morning thoroughly but in a different way. Futurity did not seem to touch her at all. What had she liked best? "Everything!" She got a kick just out of being there. Later in the day, a thought popped in my mind and I shared it with her: "Cynth, suppose you always stayed just the way you are now, never got older, never grew up. What would that be like?" Quite properly, Cynthia pointed out that she was, in fact, going to grow up. I agreed, but repeated the "What *if* . . . ?" She beamed. "That would be great!" Enthusiastically, Cynthia explained how famous she would be as "the only kid who stays a kid." She mimicked a bored celebrity signing autographs for excited little kids. Returning to her own self, Cynthia elaborated that "It would be fun! I would play all the time, all the day. I would *like* to be a kid person. *Or* a puppy! Not a dog. A puppy, a puppy, *or* a kitten! A kid person, or a kid puppy, or kid mousie, or a kid anything! I would have fun all the time, small and happy!"

I wondered if Cynthia would miss anything by not growing up. This whole line of questioning had been touched off by the contrast between David's excitement about the future and her seeming total involvement in the present. No, she wouldn't miss anything if she stayed a kid person all the time. Her explanation was incisive: "I could *play* grown-up all the time—just like I do now! Like, I play teacher. But I don't really *want* to be a teacher! It is just fun to play."

Cynthia then sighed and added, with a note of world-weariness in her voice, "You see, Daddy, a grown-up person just can't resist having a houseful of kids. I would probably want that too, I mean, when I really grow up. But if I can stay a kid person, I can play being a mommy and play being anything, and I'd never have to work really hard or anything." I wondered what she would think about if she did become a mommy someday. She replied in a tone of mock exasperation, "Who would have time to think, with all those noisy kids around!" Cynthia giggled, then turned wistful: "Oh, I would think what fun they are having. And how nice it would be if I could be a kid again. And anyhow—that's what I still am!" End of conversation.

The seven-year-old's past as well as her future is represented in the cognitive frameworks of a number of other people. Some of us, in fact, remember things about her that she does not recall. We are also likely to think about her future more extensively or, at least, in different ways then she is inclined to herself. This contrasts with the Last Leaf, whose own past, as well as the past of former friends and relations, is retained only within her personal framework. There are multiple perspectives on Cynthia's past, but the Last Leaf is alone in her recollections. This distinction contributes to the social integration of the one and the alienation of the other.

Their time perspectives differ greatly in quality and detail, in keeping with their varying life experiences and levels of maturation. But we notice that both are capable of internal maneuvers with time. The New Blossom is not too young and the Last Leaf is not too old to play with time. Cynthia often can be observed acting as if she already inhabited the future. This is a future defined in terms of her own trajectory, as distinguished from a schedule set by the calendar or other external timetable. We also know from observation that she enjoys playing "young" as well. She will be either the baby or the mommy. In other words, the girl can pivot back and forth in fantasy time. What is more, usually she can distinguish clearly between a fantasy projection into past or future and realistic recall of the past or probabilistic scanning of the future.

By comparison, the Last Leaf obviously has a more extensive past to draw upon and a foreshortened future to anticipate. Yet she may be engrossed in futuristics within her inner world, somewhat parallel to the child's use of the past. An outsider might not realize that this old person is "turned on" to the future and this child is "living in the past" at a particular moment. The Last Leaf may, for example, locate herself back in a situation in which big events were starting to unfold. She is anticipating the future again and deriving some benefits from this experience. Or she may be projecting a past scene ahead, even reconstructing the past the way it should or might have been and giving it a new setting within a future that she does not actually expect to occur. Old people may project scenes of reunion with loved ones from the past—but scenes in which the reality–unreality distinction does not seem especially critical. The individual is not confusing wish for reality; the distinction itself simply is not as important as the vision or the experience. The future projection has its own kind of reality value *as* a future projection. When I have pressed for a reality judgment, the elder has indicated that he or she couldn't say for sure. The Last Leaf, for example, does not expect to see her parents and brothers again in the same way that she expects to see the soap operas on television tomorrow. She is not entirely sold on the notion of an afterlife. But, on the other hand, she does find it natural to experience events in her own mind that have to be located at *some place* within *some* time framework. Sometimes the past will do, but at other times, the experience seems to require projection into some kind of future. It is a sort of playing with time, although no less serious for being play. We miss a lot if we pay attention only to the reality–unreality dimension and not to the quality and function of the past and future interpenetrations themselves.

2. Retrospective Modalities in Later Life

We have been touching upon some of the ways in which older people use past and future. Let us focus upon retrospective modalities for a moment. Experiences with elderly men and women have suggested to me that there are five modalities of particular importance.

2.1. The Life Review

The life review was described by Robert Butler a few years ago (1963). In his words, this is

> a naturally-occurring universal mental process characterized by the progressive return to consciousness of past experiences, and, particularly, the resurgence of unresolved conflicts; simultaneously, and normally, these revived experiences and conflicts are surveyed and reintegrated . . . this process is prompted by the realization of approaching dissolution and death, and by the inability to maintain one's sense of personal invulnerability. Although the process is initiated internally by the perception of approaching death, it is further shaped by contemporaneous experiences and its nature and outcome are affected by the lifelong unfolding of character. (p. 66)

Recognition of this process helps us to get beyond the stereotype that old people simply "live in the past." It is an active mental operation that Butler described. I think that Butler's description is sound, but I do not find evidence that the life review is universal or even that it is very prevalent. Some older men and women seem to engage in this process, and it is important for them. I suspect that we go beyond the evidence available if we assume that all old people engage in the life review.

The other retrospective modalities I offer without attempting to specify their frequency or the conditions under which they are most likely to occur; these are for subsequent research to clarify. More important for our purposes are the processes themselves.

2.2. Validation

Validation is a way of using the past to bolster the individual's sense of competency in the present. The old person scans the past for evidence that he once was competent, once was loved, once commanded respect. When this search is successful, he is better fortified to confront the challenges of the present situation: "I once was somebody—a real person, just like you." Sometimes there is nothing in the old person's life to organize a core of confidence and self-esteem around except for selected memories. You will notice that although this is a retrospective modality it has functional implications for present and future—and although it is a private, subjective orientation, it is also one that could be shared with anybody who would be willing to help the person evaluate and confirm his past and continuing value. The old person who must rely upon validation as a primary technique for meeting challenges of the day and who has nobody with whom to share this process carries a burden similar in some respects to the Last Leaf's.

2.3. Boundary Setting

Boundary setting is another retrospective modality with significant implications for past and future. The task here is for the old person to determine whether a particular place, person, ability, or whatever can still be considered part of his life or must instead be consigned to the past: "Is this part of me over and done with, or can it be carried forward in time?" Entrance into age-segregated housing, for example, and retirement are among the situations that raise questions about what in the individual's total identity and life style must be left behind, fenced off as an area one can return to only in thought.

2.4. Perpetuation of the Past

Perpetuation of the past is a term that I have lifted from the anthropologists. In this modality, the past is treated as though it were the present. Time has not really passed at all. Depending upon how we care to look at it, this modality is either the deepest submersion into the past or it is no past orientation at all. The Last Leaf perpetuates the past, but so do others, either in specific and subtle or in pervasive ways. The person who is engaged frequently or extensively in perpetuating the past runs a risk of alienating himself from the current scene, but we sometimes find that it is society's insensitivity and rejection that has placed the burden of conservation of the past upon solitary individuals instead of a consensual level of integration between past and present.

2.5. Replaying

Replaying is a familiar retrospective modality, yet its function has seldom been explored. The old person winds up to deliver, and we groan, "Do I have to hear that old story again?" If I am not totally mistaken, however, there is much that would repay our interest in replaying. Selected past experiences are taken out of their original time frames and repeated again and again. The time sequence in which they occurred is no longer crucial. The replayed experience has become, in a sense, liberated from time. It is tempting to regard this as a regressive or pathological phenomenon. Time has broken down, fragmented. The old man can't keep it together anymore. But there is an alternative interpretation. Objective, technical, businesslike time has already served its purpose. The old man doesn't have much use for it anymore. As a young man hustling for success and as a middle-aged person caught in a network of mutual obligations, he had to keep track of time. The differences between past, present, and future were important then, and the scheduling and coordination of events in time were a constant concern. But now, what time has brought forth is more significant than time itself. Time has been supplanted by "times" or "time scenes," if you will. The person is both the artist who selects choice scenes from the spectrum of times remembered and the appreciative audience whose experiences conserve what is too precious to lose.

The relationship to death is particularly interesting. Many people think of death as a "something" or a "nothing" that lies ahead—that would be in the future, of course, or at "the end of the future." If futurity itself becomes less salient, so does death. One does not have to deny death, but its meaning is altered. Replaying significant experiences in a time-free realm affords protection from the steady march "from here to eternity." The replayer, in a sense, already has a share of eternity.

Replaying does not necessarily mean that he is out of contact with passing time. I have known old men and women who show keen awareness of consensual and objective time when there is some good reason to be aware—but who *prefer* replaying and other retrospective modalities to constant immersion in the trivial, yet menacing ticks and tocks of daily life.

3. A Few Puzzles to Grow On

"In conclusion . . ." That is the way this final section ought to begin. But what I have to say is far from conclusive; it is just a few words that represent some of the puzzles I hope to grow on.

First, I wonder about the internal complexities of time maneuvering within the thoughts and feelings of a particular individual. We see that a seven-(and-a-half)-year-old girl can project herself ahead into a future time-scene in which she is thinking back upon the past—a "past" that she clearly recognizes as her immediate present. She is able to pivot forward and backward from her current position in time and do so in either "realistic" or "fantasy" modalities. The Last Leaf has less flexibility in one direction, the realistic projection of futurity, but an extensive set of possibilities for internal maneuvering within and between time perspectives. How can we somehow represent the internal structure of time perspective both as scientists and as "appreciators"? If we cannot represent the dynamic and subtle quality of internal time-maneuvers, all the interpenetrations among realms of temporality, then perhaps we are better off keeping out of other people's heads in the first place. Dimensionalizing time perspective can become an exercise in pinning and dissecting butterflies on a board. I have done my share of pinning and dissecting and perhaps will do more, but I search now for concepts and methods that will allow me to appreciate the winged creatures in their free flight.

Next, I wonder about the *interpersonal network* of time perspectives. Am I mistaken here, or are we missing an entire level of concept and method? We usually concentrate upon individual time perspectives; some of us also fashion generalizations about broad national, ethnic, or socioeconomic-echelon perspectives. But I do not see many guidelines for understanding how intimately related people enter into each other's perspectives or collectively develop, share, and modify a perspective.

And I wonder about our developmental relationship to time. Supposedly, the young child is not yet ready to appreciate time as it "really" is, and the aged person may become confused and lose this orientation, as may other people in special states of consciousness. But what we might call "establishment time" is not equally functional in all contexts and at all developmental levels. In some circumstances, it *is* functional to make fussy distinctions about past, present, and future and to keep our eye on the relentless movement of time from past through future. But in other contexts, this orientation can impede insight or enjoyment. The child and the old person may have time orientations that are more functional to them. Establishment time, in other words, may be the dominant reality within a certain range of developmental levels and situations but a bias or tyranny if imposed in other situations. In this view, it is not necessarily the case that we gradually approach and then gradually lose a hold on "real" time throughout our developmental careers but, rather, that the realities of time shift in priority and saliency with our own developmental and situational moves.

There is another sense in which I wonder about our developmental relationship to time. I see energetic, spontaneous children living intently in the moment-by-moment jostle of life. At the other end of the age spectrum, I see old men and women living in thought patterns woven from the past moments experienced. Must I choose between one orientation or the other or put both aside in favor of some hypothetical midpoint? Or is it, perhaps, that the development of a completed person and the maintenance of a resourceful society require a progression from multiple encounters with the "raw moment" to the conservation and appreciation of great-patterned time? Society needs the input of energy, the risk taking, the innovations of the young. But the developmental goal, if there is one, is to create a *completed person*. This person no longer must hurl himself into the swirling stream of the moment but can recline on the bank that rises beyond the stream. I would like to know much more about the process by which the individual moves from immersion in the moment to become a master of time rituals in which the meanings of families, generations, and nations are preserved and transformed. I would like to know how both orientations, the perspectives and the engrossments, can be understood and appreciated in proper balance. And I would like to know how to incorporate into this already-challenging problem some of the observations that have been shared in this paper. Obviously, I cannot continue to maintain that the young live primarily in the moment when I witness David and Cynthia working so actively with both past and future. And I cannot maintain that the aged function only within patterned perspectives when I witness a variety of old people who are experiencing vivid moments that just do not happen to exist at this moment in consensual time. I am not willing to simplify the complexities, the richness of minds maneuvering with time, and yet I find no easy, no satisfactory way of integrating all these frameworks and maneuvers within a single encompassing perspective.

But maybe I have drafted this kind of paper for today more for affective than cognitive reasons. Exploring the Fish Pier in Boston with my family, I had a feeling difficult to put into words, the strong sense that this morning together was an occasion *in* time that should not be lost *to* time. I found myself looking at the expressions on their faces, enjoying their characteristic interactions, loving them. What was actually going on was more or less "ordinary," and this made the situation especially poignant. The needs I felt appeared contradictory and unrealistic. Perhaps they could be expressed like this:

1. Help me to forget my sense of past and future and simply to enjoy these moments as they come and as they go.

2. No: preserve these moments in reality. Especially: let the children remain as they are, enraptured in their own ways with past, present, or future.

3. But no: Preserve these moments symbolically. Take a photograph. Make a tape. Develop a mental image that will endure through the years.

4. Or, perhaps, find a perspective in which it is acceptable for what is good and beautiful to disappear, for the ordinary to become the rare and, finally, the vanished. Construct a framework that is more secure and dependable than the stuff of life itself.

Or, to say it in still another way, "How can I appreciate this moment *enough,* without interfering, without artifice, without self-deceit?"

Not scientific, these propositions, but perhaps just another sampling of a person interpenetrated by time even before he is quite old enough to be entitled to have memories of tomorrow.

References

Butler, R. N. The life review. *Psychiatry,* 1963, *26,* 65–76.

Dali, S. *Diary of a genius.* New York: Doubleday, 1965.

Kastenbaum, R. Fertility and the fear of death. *Journal of Social Issues,* 1974, *30,* 63–78.

Editors' Introduction

This chapter critically examines some underlying assumptions that have influenced many prevalent psychological conceptions and misconceptions of time and temporal experience. It discusses the problems and probable futility of the reductionistic search for underlying "biological clocks" as explanations of temporal awareness. It questions the appropriateness of psychophysical research models and the atomistic assumptions of traditional laboratory studies of "microtime," which have largely proved to be contradictory and inconclusive. Rather, an alternative view of time as complex and shifting sets of situationally relevant cognitive-construct systems is necessary, one that emphasizes the significance of "lived" personal and social time. This view recognizes the genesis of temporal concepts and perspectives in human development. It considers the varied nature of images and representations of temporal experience and their personal and social significance. The chapter discusses how shifting cultural and social values may have biased the assumptions and conclusions of psychological research on time. The functional utility of temporal orientations in human experiences and reactions is examined, and temporal experiences in pathological conditions are considered from a functional viewpoint. Through recognition of the problems and pitfalls of earlier psychological approaches to time, the chapter attempts to suggest a more inclusive framework for understanding human temporal experience.

Images, Values, and Concepts of Time in Psychological Research

Bernard S. Gorman and Alden E. Wessman

Some years back we began an inquiry that sometimes proved rewarding, yet more often has filled us with dissatisfaction, uncertainty, and a tormenting sense of the many questions still to be asked and answered. Our quest was one that has lured many others throughout the ages, namely, a search to understand the meaning of time and its relationships to other aspects of behavior and experience.

Naïvely, we felt that the tasks of temporal research would be easy. After all, we felt, though there might be some elusive features, basically we felt that "time" was something that we knew quite well. Although time evidently had not been studied as fully as it deserved, we felt that it should not provide any particularly difficult problems. Each of us independently had already published papers on time experience and personality traits that had some promising findings, and it seemed natural to continue this course and expand the scope of investigation. As we talked to our colleagues about our plans to study temporal experiences, we found out that nearly everyone that we spoke to had occasionally thought about time but had not actively researched it because they were actively involved in other areas. A few said that they had made some

Bernard S. Gorman • Center for Research in Cognition and Affect, City University of New York, New York. Present address: Department of Psychology, Nassau Community College, Garden City, New York. **Alden E. Wessman** • Department of Psychology, City College, and Center for Research in Cognition and Affect, City University of New York, New York.

attempts in the past but were not very satisfied with their accomplishments. Most agreed that it was "about time" to start doing more and
better research on this important and intriguing topic.

We formed a very active discussion group and research team and
soon were seeing about 100 subjects regularly each week, gathering
quantities of data on the relationships between measures of temporal
experience and personality. The research questions seemed interesting
and worthwhile, and the methods seemed promising. While we found
many significant results and published several papers resulting from the
project, we gradually began to question whether we were adequately
studying what we had planned to investigate. We asked ourselves
whether we were getting any closer to really understanding the meaning
of time in human experience, and we also asked whether other
researchers had answered our questions any better than we had.

We saw two choices: either we could continue and run more
research studies, or we could spend some time in attempting to build a
better conceptual base for understanding time that, hopefully, would
lead to more fruitful research. As can be guessed, we chose the latter
course because we felt that the mere accumulation of data without a
sound theoretical framework would not be ultimately useful.

In the following sections of this chapter, we discuss some of the
conceptual and methodological issues we encountered in our reevalua-
tion of time research. Where possible, we suggest concepts and methods
that may lead investigators closer to finding important relationships
between time experiences and personality characteristics. As we
demonstrate, many of the conceptual problems with which we have
struggled are embedded in the historical development of psychology and
related fields and reflect the various paradigms that have molded these
disciplines. We believe that by critically examining some of these
approaches, we can capitalize on past victories and hopefully avoid past
mistakes.

1. Human Time and Clock Time

Over 1,500 years ago, St. Augustine asked:

> For what is time? Who is able easily and briefly to explain that? Who is
> able so much as in thought to comprehend it, so as to express himself con-
> cerning it? And yet what in our usual discourse do we more familiarly and
> knowingly make mention of than time? And surely, we understand it well

enough, when we speak of it: we understand it also when in speaking with another we hear it named. What is time then? If nobody asks me, I know: but if I were desirous to explain it to one that should ask me, plainly I know not. Boldly for all this dare I affirm myself to know thus much; that if nothing were passing, there would be no past time: and if nothing were coming, there should be no time to come: and if nothing were, there should be no present time. (*Confessions,* Book 11, Chapter 14)

Today, we are still asking the same question. We might, perhaps, ask whether time has any objective existence. Is time something that is really "out there" and a property of the environment or is time rather something that is "inside" us and, therefore, essentially a property of the person? If, instead, it is basically relational—an inner reflection and representation of external happenings—how is it possibly constructed and how closely do the two aspects correspond? Are our subjective notions of time fallible copies of some "truer" and more reliable "real" clock? As we discuss later, many experiments within the past century have attempted to see how accurately subjective human time matches the time of mechanical and electronic clocks. Before we examine that line of research, however, let us look at the notion of clocks as time-keepers.

Scholars who have studied the history and development of time concepts (e.g., Brandon, 1965; Fraser, 1975; Sorokin, 1943; Toulmin, 1965) have noted that as civilizations developed, it was recognized that subjective impressions of time based upon intuition and biological rhythms were not very reliable. As we can ourselves observe, it is possible to judge the time of day by internal cues, such as body rhythms (e.g., hunger, thirst, elimination, sleep, and attention cycles), or by external environmental cues, such as the positions of the sun, the moon, and the stars or the presence of shadows. However, these judgments are, at best, quite crude. For someone who is well-fed or for someone who is extremely hungry, time judgments based on digestive rhythms can be quite distorted. We all probably know of days when we were so excited and aroused that we forgot about meals and also forgot about time. We know that a stormy and cloudy day or a moonless night robs us of the opportunities to judge time by environmental cues alone.

Nearly all advanced civilizations adopted and invented various methods and devices that monitored time reliably and that could serve as clocks and calendars. The first clocks and calendars, as far as we know, were typically developed and used by religious bodies for the purpose of reminding people of the prescribed times for worship and for secular activities. They also served as natural or mechanical models of the presumed order of the universe. In many parts of the world today, the day is still marked by the tolling of gongs and church bells, calls to worship, and observance of the religious rituals allocated to each part of the day. Once equipped with reliable time-marking devices, a request such as "Meet me tomorrow at three o'clock" was not hopelessly subject to the whims and fallibilities of each person. Instead, plans and schedules could be effectively executed if each person used the same kind of clock. As civilizations expanded their geographical boundaries and increased in complexity of social organization, more accurate clocks and calendars were needed for navigation, communication, and commercial purposes (Sorokin, 1943).

Today, it is possible to buy an inexpensive electronic watch with an accuracy of ±1 minute per year. While these watches please their owners by giving them a sense of precise, technical mastery in an otherwise chaotic world, they generally have little practical advantage for the average person, for whom a delay or advance of a few minutes probably makes little difference in most aspects of his social or occupational roles. However, for the aircraft pilot, the broadcaster, and others in highly technical roles, a watch that loses a few seconds a day may not be accurate enough.

As we can see, a reliable and orderly clock often gives us a sense (albeit illusory) of an ordered and reliable social and physical world. Some people find comfort in such order, as it establishes the very boundaries of their existence, while others feel terrified and constrained by such limitations and seek solace in romantic and mystical "escapes" from a predictable time scheme. At this point, we might wonder whether our search for the meaning of time has ended. Perhaps, we might simply define time as "what the clock measures." To do so would be premature as we would be granting a reality to clock time that is unwarranted.

In choosing any particular clock as "our" clock, we essentially agree that an arbitrary device will help us collectively to form more reliable judgments of social and technical events than each of us can make individually. But if we examine our fiat, we will see that our clock (or any other clock) is no more the "essence" of time than rulers are the "essence" of space. They are merely conventional measuring devices. As v:e shall see in further sections, the equation of all times to clock time has produced serious difficulties.

2. Psychophysics and the "Sense" of Time

The fact that human beings can be regarded as complicated electrical, chemical, and biological "mechanisms" and, at the same time, as consciously experiencing, feeling, willing, and transcendant "beings" has provided the major riddle of human existence. We know that before conception and after death, the human organism can be adequately described in physiochemical terms. Between these two limits, however, a mysterious process called "life" intervenes, and the organism grows and reproduces and, more importantly, also possesses an awareness of its own existence. Attempts to solve the puzzle of the meaning of life have produced many secular and religious cosmologies that have attempted to reconcile the fact that at some point inanimate matter becomes the "living" and eventually returns to the nonliving again.

In a fascinating account of the history of clocks, Fraser (1975) has noted that originally the robotlike mechanical figures that struck chimes and marched about early clocks were not meant simply to be the droll, decorative figures of cuckoo clocks. Instead, these figures served as mechanical models of the life process in much the same way modern scientists build computer or mathematical models of behavior. Not only were the early clocks modeled after life, but, as we sometimes see in psychopathology, the conceptual relationship can be reversed so that the life process is sometimes conceived as being modeled after the clock!

René Descartes (1596–1650), a seminal figure in psychology, mathematics, and physics, was puzzled by the duality of human nature. Descartes proposed that the mind (or the spirit or the soul) and the body were separate entities that somehow interacted. For Descartes, animals were automatons—mechanical forms without minds—much like the clocks of his day. Man, however, had, in addition to his corporeal existence, *res cogitans* or "thinking matter." In Descartes' system, the pineal gland was cited as the locus of interaction between mind and body. Although Descartes obviously did not provide the ultimate solution to the mind–body problem, he introduced the notion of dualism. Succeeding schools of thought in philosophy and psychology have broached this problem by sometimes accepting a dualism and by sometimes adopting various forms of monism, the belief that mind and body are both aspects of one totality.

Modern psychology emerged from the confluence of biology, physics, and philosophy in the late 19th century as a form of experimental "natural philosophy." The early psychologists accepted a form of dualism and sought methods that would enable them to find a correspondence between the mental and the physical realms. For this purpose, the psychophysical methods developed by E. H. Weber and G. Fechner in the 1860s were rapidly adopted. Psychophysics attempted to discover quantitative relationships between the physical properties of stimuli and the reported experiences of persons exposed to the stimuli. Because physical stimuli seem to have an objectively measurable existence beyond that of an individual observer and can be assessed in energetic or spatial units, and because conscious reports of stimuli have a highly "mental" nature, the psychophysical methods tend to support a dualistic or interactionistic viewpoint.

Review articles of early experimental research on time appeared in the new psychological books and journals (e.g., James, 1890; Mach, 1865; Nichols, 1890) and revealed an extensive use of the same psychophysical methods that were being employed in studies of vision, audition, and other senses. In these methods, the subjects were typically asked to estimate or reproduce or produce intervals of clock time. Mathematical functions that expressed relationships among subjective time estimates and clock times were then derived.

While the findings of psychophysical research have tremendously aided our understanding of sensory processes, we seriously question their appropriateness and utility for helping us to understand time. Let us remember that in a traditional psychophysical study (as in visual or acoustic research), the subjects are presented with some objectively quantifiable *physical* stimulus (e.g., a light source with a given wavelength, amplitude, and saturation) and then are asked to form some subjective judgment of the stimulus characteristics. However, time studies cannot fit this paradigm, as "time" is not a simple, direct physical stimulus to begin with, and clock time is, as we have previously seen, an arbitrary system that aids observers in coordinating and ordering events. Although we can perform "psychophysical" studies of time, the product of such research tells us not how subjects respond to a basic physical stimulus but how accurately they can *translate* their personal experience and judgment of events into the more publicly-held standardized system of judging events. While translation functions may be useful in investigations of how much and under what conditions an individual's judgments deviate from the "correct" clock-time standard, we are actually several steps removed from the traditional use of psychophysics.

If temporal judgments are not basic sensory judgments, what are they? As such, temporal judgments are highly individualized estimates employing various internal and external cues in conjunction with cognitive constructs concerning the order, the duration, and the rhythmic nature of events (Doob, 1971, pp. 30–46, 152–190). As many of the authors in this volume have shown, time is a set of acquired mental schemata and concepts that individuals and cultures have developed for various purposes. Time is not an external "object." Although we may commonly use *time* as a noun in everyday language, we probably would find it more advantageous to consider time as an internal *process of judgment* rather than a primary object of judgment itself. St. Augustine recognized this long ago, when he wrote: "It is in thee, my mind, that I measure times" (Whitrow, 1961, p. 48). And in his monumental *Phenomenology of Perception,* Merleau-Ponty (1962) placed the locus of time in the subject rather than in the environment, when he said: "Time is, therefore, not a real process which I am content to record. It arises from *my* relationship to things" (p. 412). Many other thinkers have recognized this, as we have seen, yet the misconception of time as an external, objective property or thing persists.

3. The Search for the "Organ" of Time

Over the past century, there have been hundreds of studies that have investigated the "time sense" by psychophysical means. As researchers were readily drawn to the metaphor of a "time sense," it seemed quite natural to search for an underlying "time organ" that would serve as a transducer or processing center for "temporal stimuli."

Several lines of evidence gave support to the tantalizing notion of a time organ. It had often been observed that when there were alterations in the structure or functions of the nervous system—as in brain lesions, traumatic injuries, toxic states, convulsive seizures, and other massive electrical, thermal, or chemical changes—there were often gross alterations in time judgments. Temporal judgments produced by impaired subjects were often so strikingly deviant from the time judgments of normal subjects that many investigators were led to postulate that a "time center" or "time organ" must be located at the sites of the injuries.

Another pathway that led to belief in a temporal organ can be found in the many studies (see Fraser, 1975, pp. 178–195; Luce, 1971; Pittendrigh, 1971) that indicated that many biological systems in both animals and plants have their own characteristic rhythms, which are, indeed, very clocklike. Among these rhythmic processes are circulation, elimination, and respiration cycles; intrinsic brain rhythms; hormone cycles; and many others. As these body processes are essentially clocklike, many researchers assumed that the rhythms could serve as reference sources for time judgments. For example, if the heart beats at a steady 72 beats per minute, then an event that occurred during a period when the heart beat 18 times would be judged as being approximately 15 seconds long. There is now substantial evidence that regular biological rhythms undoubtedly exist, but whether they actually do serve as clocks and time bases is doubtful.

Several factors prompt us to abandon biological time organs as necessary and sufficient explanations of human temporal experience. In fact, we may find that the search for separate time organs or biological clocks is unnecessarily reductionistic. No one would deny that the brain is involved in a huge number of perceptual, motoric, and cognitive functions. However, the assertion that a specific nervous system site is responsible for highly complex time judgments is fairly naïve. For example, we know that complex behaviors typically involve many interacting brain regions and that a given brain center, while having somewhat autonomous structures and functions, probably never works in isolation except in the rarest of pathological conditions (Goldstein, 1939; Lashley, 1929). The idea of completely autonomous centers of the brain probably developed from beliefs in phrenology, which postulated that certain numbered regions of the skull were related to character traits and that people would display different personality patterns related to their physiognomy.

It is fairly naïve to assume that particular behaviors and experiences are directly correlated with and are isomorphic to specific brain centers. Even the relatively good mappings of visual functions in the occipital lobe and motor functions in the pre-Rolandic areas of the brain are not sufficient to explain visual perception and motor activity. We would not expect, for example, to find a "beauty center" in the brain simply because people tend to consider other people or things as "beautiful." Neither would we call the elbows or kneecaps the "organs of locomotion" because they are instrumental in walking and crawling. However, some researchers have persisted in searching for temporal "centers" because people make temporal judgments.

A very strong argument against the "time organ" approach was provided by Ornstein (1969), who stated: "The argument is not that increases in body temperature (or in the speeding up of a 'biological clock' with a drug) do not lengthen time experience, but rather that these manipulations are more parsimoniously considered as affecting cognitive processing rather than altering one of the maze of possible 'chronometers'; heart rate, tapping rate, body temperature, cell metabolism, breath rate, etc." (p. 34). Thus, Ornstein indicated that the search for the biological clock is unnecessarily reductionistic and that time judgments are best studied at the level of cognitive systems rather than at the level of biological substrates.

Another problem inherent in the biological clock approach can be found in the question of whether biological rhythms could be the causes of temporal judgments or whether they themselves might be correlates or effects of cognitive processes. An interesting study that bears on this issue was performed by Jenner, Goodwin, Sheridan, Tauber, and Lobban (1968), who observed a manic–depressive patient who displayed cycles of mood and activity in which he was elated for a 24-hour period and then depressed and lethargic for a 24-hour period. The patient had displayed this 48-hour cycle for an 11-year period, and measures and biological assays of electrolytes, 17-hydroxy ketosteroids, pH, creatinine, and other physiological parameters also followed the 48-hour cycle. However, the experimenters wondered about what would happen if the patient was isolated from his usual surroundings and was presented with a speeded-up daily schedule in which food schedules and lighting conditions were altered so that 11 real days equal 12 experimental days. It was found that under this new regimen, the patient's moods and biological indicators both rapidly shifted to a new 44-hour cycle and therefore reflected the rhythm of his surroundings. The study indicated that the biological clock could, perhaps, be as much a product of cognitions of time and events as it could be the cause of cognitions of time. Whether such rhythms can be altered indefinitely is open to question, but the results cast doubt upon a simple biological clock explanation of time experience.

Finally, we must ask whether we ordinarily bother to use the data provided by biological rhythms for our time judgments. For example, Luce's book *Body Time* (1971) indicates that among the many rhythms of the body, there is even a nostril rhythm. A person usually breathes through one nostril for about three hours while the other nostril is slightly engorged. Then the process is reversed and the previously engorged nostril opens and becomes the dominant nostril for breathing within the next three hours. If one were particularly obsessional, possibly he could time events using the nostril rhythm. However, it is unlikely that anyone would consciously or even unconsciously base time judgments primarily on this single esoteric rhythm or even on any of the other more salient bodily rhythms. Though our "sense of time" based on internal physiological cues and sensory information processing may adequately serve for rough approximations, we recognize that our autochthonous judgments of time are quite unreliable. When we are really concerned about time, we consult the mechanical clocks so readily available in this society (Doob, 1971, pp. 152–154).

4. Atomism and Experimental Studies of "Microtime"

Many of the early efforts in time research viewed time as sensory and perceptual in nature, and specific organs or functions were postulated as constituting the basis of the time sense. The psychophysical experiment clearly gave an impressive aura of scientific respectability to time research. Who, for example, could accuse a researcher of not being "scientific" when, after all, he was surrounded by instrumentation that could present stimuli and record response data with split-second accuracy? The very act of painstakingly collecting reams of data on judgments of the most minuscule time intervals seemed to attest to the validity of the notion that the accurate collection of findings on the experience of short time intervals would ultimately provide important information about the judgment of longer time intervals. The belief that discovering the elementary building blocks or components of phenomena would eventually lead to an understanding of larger and more complex phenomena is known as the doctrine of *atomism*. Although this doctrine was long ago rejected by the functionalist and Gestalt psychologists and field theorists, and by ecologists in the physical and natural sciences, traces of atomism still persist in time research. At least this is our

assessment of much of the dreary and inconclusive experimental literature so painstakingly reviewed by Fraisse (1963) and Doob (1971).

For time researchers, one expression of atomistic beliefs took the
form of conducting research in which attempts were made to extend the
findings gained from laboratory studies on judgments of short time
intervals to explanations of the experience of larger segments of time.
The implicit assumption seemed to be that if clock time intervals were
linear and additive, then the subjective experiences of time should be
similarly additive. Unfortunately, much research has shown that this
form of atomism has not been particularly productive.

It is now fairly well established that the ability to judge short time
intervals is insignificantly correlated with the ability to judge longer
intervals (Gilliland, Hofeld, & Eckstrand, 1946; Loehlin, 1959), and
therefore a simple additive process cannot be assumed. Doob (1971)
pointed out that the very act of forming a time judgment is never really
simple and probably always involves a complex set of "secondary" judgments in which each subject weighs, combines, and categorizes large
amounts of information inherent in the total experimental setting. In
essential agreement with Doob's notion of the judgmental nature of
time, Gilliland *et al.,* (1946) stated:

> Time estimation is not nearly as given from sense data as in space percep-
> tion and estimation. Time estimation partakes more of the nature of a
> judgment rather than a perceptual process. The fact that in space percep-
> tion the cues remain for further examination whereas in time perception
> after the event it is always in retrospect, helps to make time estimation
> more of a process of judgment than perception (p. 173).

Perhaps there are no elemental experiences of time, and if there are
any, it is difficult, if not impossible, to observe them in a laboratory setting. So far, the reviews of the past century's research (e.g., Doob,
1971; Gilliland *et al.,* 1946; Nichols, 1890; Weber, 1933) have not
revealed any consensus on elementary time experiences.

5. From "Laboratory Time" to "Lived Time"

As we have seen, the laboratory investigations of subjective temporal judgment have not demonstrated any consistent relationships
between judgments of shorter and longer time intervals. The findings
are also greatly dependent on the particular experimental method used.

While long intervals of clock time are direct mathematical multiples of shorter clock intervals, the same relationship does not seem to hold for experiential time.

A far more serious limitation of traditional laboratory research, however, seems to be in the lack of relationships between the artificially constricted laboratory studies of "microtime" and the personally relevant and meaningful experiences of time that have often been called "lived time" by the existentialists. For example, how does the knowledge of how accurately a subject judges an interval of a second, a minute, or an hour help us in understanding how the same person will plan his future, reminisce about his past, or experience his present? In the realm of "lived time," we would expect to find rich and complex variables that are very different from the limited psychophysical judgments of simultaneity and duration.

Much of the literature on lived personal time has been concerned with "time perspective," which can be roughly defined as the degree to which a person, group, or society conceptualizes events removed from the present situation. Time perspective may, in turn, have to take into account further aspects such as the evaluation of the past, the present, and the future. For example, one individual might be highly concerned with the future and horrified by its possibilities, while another might also be highly concerned with the future but view it as the fulfillment of many hopes and promises. We may also wish to examine the relative degree to which a person allocates his thoughts to the "zones" of past, present, and future. Most probably, no individual is entirely oriented to the future or the past or the present but has, instead, distributed his thoughts among the events in each time zone.

Another important concern of researchers who have studied lived personal time can be found in their emphasis on the varied meanings and uses of time symbolization. As we discuss later, constructs of time are often represented and concretized by a variety of verbal and graphic symbols and images. By discovering why and under what circumstances individuals and cultures choose one form of temporal symbol or metaphor over others, we may be able to learn something about the people who construct and employ the symbols. For example, the symbolization of time in different psychological disorders may provide insight into the phenomenological world of the people experiencing the disorders (Ellenberger, 1958; Straus, 1966, pp. 290–295).

A search for the meaning and the functions of lived time should also examine the development of time concepts in the individual. Such an examination would help us to understand when and how the person's views on time evolve during the life span.

In general, lived time research will be strongly concerned with individual differences in temporal experience and concepts. In addition to inheriting the problems of having to define the nature of "time" in its various aspects, such individual differences research has the additional burden of having to explain the factors that account for both unique and systematic differences among people. An important focus of such research is its emphasis on the personal meaning and significance of temporal experience. In the following sections, we discuss some of these problems as they apply to selected aspects of time research. We review some of the important work that has been done and present our views regarding what might be undertaken.

6. Research on the Genesis of Temporal Awareness and Concepts

Considering the wealth of research on child development and the continuing curiosity about the nature of time, surprisingly little good empirical research and theoretical formulation has been published concerning the development of time in children (Wallace & Rabin, 1960, pp. 213–216; Doob, 1971, pp. 213–222). Among the major schools of thought in developmental psychology, we find the *psychoanalytic school,* the *cognitive developmentalists,* and the *behaviorists.* Of the three major schools of thought, behaviorism has produced the least research on the subject of time. This is not particularly surprising, as behaviorism has generally avoided any consideration of private internal events such as thoughts, intentions, judgments, and feelings, as useful sources of data, preferring external observations of overt behavior rather than analysis of internal cognitive and affective experience. Aside from some discussions of response latency, delay of reinforcement, interstimulus intervals, and differential schedules of reinforcement, little has been said concerning time by the behaviorists and, as far as we know, nothing has been said regarding the subjective experience of time. Psychoanalytic and cognitive developmental approaches, however, have made some important contributions regarding the development of temporal awareness, attitudes, and concepts.

According to Freudian psychoanalysis (Abraham, 1927; Bonaparte, 1940; Freud, 1960; Fenichel, 1945; Bergler & Roheim, 1946), the infant is originally equipped with the *id,* an undifferentiated and unstructured set of impulses, affects, and primitive cognitions that operates without reality constraints. This primitive system functions reflexively and unconsciously and seeks immediate body pleasures without regard for appropriate times, places, or modes of gratification. During the oral, anal, and phallic stages of psychosexual development, the child's wishes for immediate gratification are frequently frustrated and are forced to accommodate to reality and the social framework. Through systematic frustration experiences, the child gradually develops the reality-oriented and socialized structures of the ego and superego, which include the ability to delay gratification, the ability to plan and schedule, and the concepts of trust and patience.

In Erikson's (1963) view, the Freudian "oral" stage is a period in which the basic conflict of "trust versus mistrust" arises. The child interacts with the mother and other significant figures during this stage and forms fundamental attitudes and rudimentary concepts concerning the trustworthiness of the people affecting his well-being. In this early stage, primitive forms of such time-related concepts as optimism, pessimism, disappointment, and hope tend to arise, and persisting representations of the self and the object world are initially established.

In the anal stage, characterized as a conflict of "autonomy versus shame and doubt" by Erikson (1963), the child's impulses are restricted and constrained by the socialization demands of a larger society. In Freudian analysis, the toilet training situation serves as the paradigm for this stage, and here, the child learns to follow schedules and to expect rewards for adherence to other's rules concerning time, place, and mode of behavior. Conversely, the child learns to expect punishments and shame for disobeying social rules. The autonomy of the self is often pitted against the demands of others in this stage, and the child soon learns that his own desires must be gratified within a social framework that extends beyond his own immediate concerns. Many psychoanalytic discussions of obsessive and compulsive behavior have implicated this stage as being crucial in forming the child's notion of schedules and punctuality.

In the phallic stage, the child is confronted with the problem of managing jealousy, especially in the family's triangular relationships of the Oedipal and Electra complexes; the discovery of sex roles and gender identification; and the internalization of self-controls through identification with other people. According to psychoanalytic theory, the concept of guilt, which includes such temporal features as anticipation and regret, is formed in this stage.

Psychoanalysis typically views later behavior as having its roots in the pregenital phases. However, if we assumed that psychoanalytic writing on time invariably attributed temporal experiences to earlier phases of development, we would be quite mistaken. On this point, the analysts Bergler and Roheim (1946, p. 197) stated that statements such as "pressed for time" and "time runs out" cannot be unambiguously traced to the anal stage. Nor can we trace such experiences as "time that devours everything" to the oral stage. Instead, Bergler and Roheim have argued that the psychoanalytic theory of time should not be taken literally as specific experiences tightly anchored to particular stages, but rather should be taken as examples of typical situations in which the child abandons egocentric modes of cognition and affect and adopts more socially structured modes of existence. Bergler and Roheim's contribution transcends the restricted formulations based on specific cultural practices in a limited historical context that characterized early psychoanalytic theory and paves the way for broader viewpoints with greater cross-cultural and historical generality.

Erikson and Bergler and Roheim have pointed to the importance of the basic *processes* of psychosocial, psychosexual, and cognitive development and of *variations* according to specific cultural contents and contexts. For example, we might be able to consider variations in the acquisition of hope and trust in societies that are orally indulgent *and* orally restrictive; the acquisition of scheduling and timing in societies that are not particularly concerned with early toilet training; and the acquisition of guilt in societies in which the family structures are very different from those of traditional American and European families. Thus far, there have been only a limited number of adequately conceived and designed cross-cultural developmental studies of temporal attitudes and orientations.

As Voyat and we have discussed in previous chapters, Piaget's (1946, 1966, 1970) formulation of the development of temporal schemata in the child is a general process-oriented theory that is not necessarily restricted to any particular cultural context, although its universality and normative generality remain to be firmly established. In Piaget's theory, all cognitive development proceeds originally from reflexive actions and concrete empirical representations to progressively more formal and abstract symbolic constructions of reality. In Piaget's theory, cognitive development is the result of a series of cycles that result in states of dynamic equilibrium between two tendencies: *assimilation,* the forcing of cognitions into preexisting schemata, and *accommodation,* the tendency to alter schemata in the light of new information imposed by observed events. Piaget views the development of temporal concepts as following a course similar to that found in the development of concepts of space, volume, and matter.

Most importantly, as Piaget (1966) and Voyat (this volume) have shown, the concepts of time are not *a priori* "givens" but are always derived by the child from relationships among more primitive concepts of effort, work, power, velocity, and distance. For Voyat and Piaget, distortions in time judgments by both children and adults are fairly common because the individual is never directly judging time but is, in fact, labeling the end result of a subjective equation as "time." Piaget (1966, p. 212) illustrated this notion of derived or constructed time by discussing the subjective equation in which *time = work/power.* Piaget noted that when one is doing a task, if power is increased, time seems to diminish. If the amount of work is increased, however, and the person's ability to do the task remains the same, time seems to increase. Piaget observed that according to this derived time equation, our subjective impressions of the length of time it would take to walk over a given distance would differ if we were walking through heavy snow or on a dry pavement. The time it takes to walk through snow appears longer because the amount of work required is greater. Voyat expanded upon this idea by demonstrating how children derive a concept of time from relationships among velocity and distance. As Piagetian research methods are amenable to cross-cultural investigations, we may be able to examine how individuals in other societies possibly form their "equations" of time from other fundamental concepts.

The strategies employed in developmental research on time concepts may present a number of problems. Among the basic research designs in developmental psychology, the cross-sectional and the longitudinal approaches are the most prominent. In the cross-sectional design, representative groups of subjects are chosen from each of several age ranges, and differences on a given task are observed. For example, a group of 5-year-olds, a group of 10-year-olds, and a group of 15-year-olds are given a rote learning task, and it is found that the older groups perform progressively better than the younger groups. Systematic differences in performance among the groups are interpreted as indicators of a growth trend.

In the longitudinal method, a single group of same-aged subjects are chosen at a certain point in their development and are repeatedly assessed on a given task as they reach different ages. In this way, the subjects serve as their own control group, and systematic differences in their performance over time are taken to be evidence of a growth trend. For example, at six-month intervals, starting from the age of 6 months until the age of 6 years, counts are made of the number of different words emitted by children within a one-hour observation period. It is found that in successive years, the number of different words in the children's vocabularies increases at a negatively accelerated monotonic rate.

Several researchers, including Buss (1974), Riegel (1969), and Wohlwill (1970), have pointed to some serious defects in the cross-sectional and the longitudinal methods and have encouraged the use of more sophisticated designs. Essentially, the problem with the usual methods lies in the fact that while an individual's behavior may change over time because of intrinsic age-related factors and acquired skills, the person's society is also changing in many of its institutions, including child-rearing practices, kinship structures, folkways and folklore, means of production of goods and services, and many other aspects of culture. In this way, marked differences other than those due purely to age differences may be found in people of different ages. Thus a 10-year-old in 1977 is different from a 40-year-old in 1977 not only by virtue of a 30-year age difference but also by the fact that the 40-year-old was born in 1937, when the society was very different from the society of 1967, when the 10-year-old was born.

The problem of separating age-related variance from generation-related variance is especially critical in time research. At the time of this writing, the authors are in their 30s and 40s and grew up in a world quite different from that of their children. We were not exposed to many things that our children easily take for granted. We spent many hours reading or listening to radio programs, while our children generally use the radio as background sounds and from an early age have been exposed to television. We traveled by foot, car, and train, and a few people traveled in propeller-driven planes. Our children, however, have never known a time without television and computers. Satellite com-

munications, space travel, and rapid jet flights are seen by them as axio-
matic aspects of their world views. Even the ways of telling time from a
circular watch or clock dial are changing as we are seeing the rapid
adoption of digital electronic timepieces.

Therefore, when we assess differences in the acquisition of time
concepts and time perspectives at different ages, we must also be aware
that we may be seeing generational effects as well. Thus, a statement of
the form "Elderly people are less oriented toward the future and more
oriented toward the past than younger people" may have to be modified
by considerations of *when* the people were "elderly" or "young." In
this volume, Riegel argues for a model of time that includes both age
and generational factors. And Kastenbaum and Cottle both show that it
is difficult, if not impossible, to make a blanket statement about age differences in time perspective without considering many other variables.
Excellent reviews of generational effects can be found in Riegel (1972),
and cogent presentations of research designs that consider age and
generational effects can be found in Buss (1973, 1974).

7. Research on the Representation of Temporal Concepts

When we express our concepts of "time," we often symbolize time
in the form of some graphic visual image or in some verbal metaphor
or musical motif. Piagetian and psychoanalytic approaches both point to
the importance of the representational nature of time. For the analyst,
time symbols are acquired through the socialization of psychosexual
drives. For the Piagetian, "time" itself is often a representation and a
metaphor derived from more basic processes. Knowledge of the meanings and functions of an individual's or a culture's repertoire of time
symbols may allow us better understanding of how basic values and perceptions are projected onto the abstract concept of "time."

We know that time can be graphically represented in many forms.
Among the most common are the images of time as a line, time as a
wheel, and time as a pendulum (Leach, in Cottle & Klineberg, 1974).
Still other visual images of time can be seen in representations of time as
shifting sands or rising and falling tides. While some cultures generally
prefer and utilize one form of symbolism over another, it is quite possible for individuals within a culture to use a large variety of symbols for
different purposes. For example, the cycle of holidays and birthdays and

the succession of generations within our own families might prompt us to view time as circular. Traveling down a long highway at a constant speed can easily enforce the notion of time as a directed line. Swings of mood and shifts in luck, however, may prompt us to view time as a pendulum. Depending upon a person's ability to construct imagery, his time representations may be quite stereotyped or quite idiosyncratic and unique.

When we elicit images of time from research subjects, we often find large differences among subjects in the use of graphic images of time. In several of our own studies, we pursued the time line image and used a simple time line technique adapted from previous studies by Cohen, Hansel, and Sylvester (1954) and Rychlak (1972). In this technique, we asked subjects to delineate certain time periods on two 10-inch lines. The first line contained the end points of "birth" and "now," while the second line contained the end points of "now" and "death." In our first study (Gorman, Wessman, Schmeidler, Thayer, & Mannucci, 1973), subjects located points on the line in a psychophysical "power law" manner, so that periods close to the "now" point were represented as disproportionately larger than the more remote points on both the past and the future line tasks. In further analysis (Wessman, Gorman, Schmeidler, Thayer, & Mannucci, 1974), we found that the more anxious, impulsive, and undisciplined subjects drew larger representations of the present than the calm, controlled, and responsible subjects. Conversely, the impulsive and anxious subjects drew smaller representations of the distant future and the distant past. Perhaps the proportionately larger representation of the present by these subjects was an indicator of the greater importance of immediate events in their lives and indicated that past memories and future anticipations were less salient in regulating their behavior.

In some of Cottle's (1967, 1969) studies and some of our own unpublished research, subjects were asked to represent the past, the present, and the future with circles. The subjects were instructed to draw the circles in any size that they preferred and were told to place the circles in any pattern that they wished. The modal pattern in our study was one in which the subjects made their future circles largest, their past circles smallest, and their present circles intermediate. Circles were usually overlapped so that the past circle touched the present circle and the present circle overlapped the future circle. The representation seemed to indicate that our college-age subjects saw their futures as the most important times of their lives and the past as the least important. The overlapping nature of their representations indicated that these subjects saw time as integrated, although Cottle's (1967) subjects generally tended to draw isolated past, present, and future circles. While we found that our subjects *could* represent time in linear or circular representations, we were not at all sure that they would spontaneously choose to represent time in these ways. In fact, we have evidence that some of the subjects felt that the representation tasks forced them into unnatural modes of representation with which they were not comfortable.

When we examine the verbal metaphors for time, we are confronted with an even larger array of possibilities than in visual representations of time. Some of the metaphors for time seem to be quite contradictory. For example, Meerloo (1966) described some metaphors for time that were used by his psychoanalytic patients. Among these metaphors were: time eating away (e.g., "ravaged by time"); time as an arrow; time as the two-faced god, Janus; time as an obsessive–compulsive repetition ("time and time again"); time as money; time as boredom; and time as creation. Knapp and Garbutt (1958) and Wessman and Ricks (1966) developed time metaphor questionnaires in which subjects were asked to indicate how closely various metaphors corresponded to their subjective notions of time. It was anticipated that different types of metaphor preferences would correspond to important personality characteristics. Knapp and Garbutt (1958), for example, found that subjects who preferred metaphors that describe time as dynamic, swift, directional movement were usually people with high needs for achievement, while those who were rather low in achievement needs tended to choose metaphors that describe time as slow-moving or completely static. In their book *Mood and Personality,* Wessman and Ricks (1966, pp. 117–120) found that happy men preferred metaphors that describe time as *ascending* (e.g., "a soaring bird," "a shooting star"); as *growth and development* ("pregnancy and birth," "an ever-branching tree"); as a *good person* ("the wisest of counselors," "a kind physician"); as *harmony and order* ("a Bach cantata," "the order of nature"); and as *active and directed* ("continuity of aim," "the thrust of forward purpose"). Conversely, the less happy men chose metaphors that describe time as *descending* ("a slow descent," "a stairway to the tomb"); *decomposing* ("a collapsing edifice," "a rotting tree trunk"); a *bad person* ("the old bald cheater," "a chronic thief"); *monotonous* ("a treadmill without end," "a tireless automaton"); *barren and empty*; and *overpowering and enslaving* ("something you can never stop," "something you are not ready for").

Literature on the phenomenology of time in severe psychopathology also provides a rich set of time metaphors. Paul Schilder (1942) discussed the case of a psychotic patient who stated, "My head is a clock, an apparatus. I make time, the new time as it should be." Another of Schilder's patients said, "Is there any future? Previously I had a future, but now it shrinks more and more." In a paper concerning the expression of time in schizophrenia, Seeman (1976) focused on the meaning of time for a schizophrenic woman in psychotherapy. The patient would make statements such as, "I have to watch, check, count, measure and plan," "I might as well give up the present; I'll live in the past," and "Everyday is a new beginning; therefore, nothing matters, nothing counts" when she felt that her attempts to gain mastery over the events of her life were futile. The patient would claim to possess clairvoyant abilities in order to predict her confusing world. Struggles for autonomy were expressed in statements such as, "My time is not my own. Someone else is always telling me what to do." While some of these disturbed representations of time are more extreme than those usually encountered, they nevertheless illustrate how the symbolization and experience of time is meaningfully connected with the person's sense of his own existence.

With such an abundance of striking symbolizations of time, should a major research endeavor be one of collecting symbols, categorizing them, and then searching for their correlates? Past research has often done this and some meaningful findings have emerged from these efforts. However, before a decision is made to adopt this tactic, it might be useful to examine the nature of time symbolization further.

One important aspect of time symbolization that we found rather surprising was the tremendous number and variety of time symbols and metaphors. These representations of time did not seem to have any fixed meaning or universal usage. As we considered this issue, we realized that the lack of fixed meaning is not only a property of time symbols but is found in all "nondiscursive" symbolic productions such as dreams, daydreams, art, and poetry (Langer, 1951, pp. 75–94). Freud (1967, Chapter 6 [D,E]) questioned whether there were any fixed symbolic meanings in dreams. He had observed that certain sexual symbols seemed to appear quite regularly in dreams. Freud, however, warned against a simple "dream book" decoding of symbols and strongly argued that the analyst should attempt to decode the meaning of the dream symbol within the subject's own frames of reference and associative schemata. We feel that Freud's warning against "dream book" decoding is also applicable to time symbolism. The lack of fixed meanings for time symbols has some important implications. We should view temporal symbols not as distorted or altered perceptions of a determinate thing called "time" but rather as representations chosen and constructed by individuals as apt expressions of their own life situations and feelings.

Viewing temporal symbols as expressions or projections of an individual's life situation and personal system of values, meanings, and feelings raises many important questions concerning the process of symbol formation and usage. We present here a number of the questions we see and indicate our thinking regarding some of the issues they raise.

One set of questions involves the matter of the "choice" or "selection" of particular types of symbolic representation and their significance. Why, for example, does a person choose one particular set of symbols out of all the possible, available symbols for time? And why do individuals or societies choose to symbolize some aspects of their lives in temporal terms? Are their time symbols consistent with a more general outlook on their lives? For example, in his discussion of character orientations, Fromm (1947) indicated that in modern capitalist societies, people may view themselves as marketable commodities and may adopt a set of symbols concordant with their perceptions of themselves. Perhaps it would not be unusual to see such expressions as "saving time," "spending time," and "buying time" as symbols not simply of time but of the person's own view of his actions and life style.

We may also wish to investigate how temporal concepts are organized by the individual. Do people use the same concepts for all aspects of their lives, or do they employ different concepts for different situations? It may be that some individuals use the same time metaphors for all of their activities, while others may tend to employ very restricted sets of symbols that differ as they work, engage in leisure, enter private activities, and take their roles in interpersonal encounters. Is the consistent use of a single representational system maladaptive in some circumstances?

We can also ask, how common or unique is an individual's system of representing time? As a person internalizes some of the values and cognitive styles of his society, he undoubtedly acquires some common time metaphors and symbols. For example, many of us have learned such homilies as "time flies" and "a stitch in time saves nine." We probably have some vivid images of clocks, calendars, and other concrete representations of time. However, we also know that people differ in the degree to which they use these representations. Can we postulate that deviance from a culturally modal pattern of behavior will also be accompanied by some deviant representation of time? Conversely, can we postulate that conformity to a culturally modal pattern of behavior will entail the adoption and internalization of conventional representations of time?

Another set of concerns involves methodological questions regarding how best to assess the public and common and the individual and unique representations of time. In their extensive research on connotative meaning, Osgood and his colleagues (Osgood, Suci, & Tannenbaum, 1957; Osgood, 1971) have shown that the major proportions of variance in the connotative meaning of words can be explained by three major factors: *evaluation, potency,* and *activity*—and, perhaps, four minor factors: *concreteness/stability, human/nonhuman, ordinariness,* and *age.* Osgood has shown that the first three major factors seem to appear in factor analyses of meaning in many different cultures.

In some of our own unpublished studies, in which we attempted to develop factor analytic scales for assessing individual differences in temporal experiences, we obtained factors such as: *optimism versus pessimism, scheduling and planning abilities, change seeking, punctuality, sad versus happy past experiences, memory of the past, boredom and disinterest, preferences for slow and leisurely pace,* and *feelings of involvement and absorption.* As we look at these factors now, it appears that some of them are strongly evaluative. Some of them reflect a sense of the individual's power to influence events in his own life, or they express feelings indicating that the person has been overpowered by time. Several factors indicate a sense of activity and pacing. It would not be very surprising to find that many of the meanings and metaphors for "time" may reflect some of Osgood's semantic differential dimensions in disguise!

In some of our planned future research, we intend to develop semantic differential scales for such concepts as the near and the distant past, the near and the distant future, and the present. We may build separate scales for personal time concepts and impersonal time concepts (e.g., "My Recent Past" versus "The Recent Past").

Osgood's semantic factors seem to be promising tools for assessing commonly held and culturally shared meanings of time concepts. However, individuals might use dimensions other than the Osgood factors. Furthermore, the dimensions may be organized quite idiosyncratically within each individual. Several techniques have been evolved for studying intraindividual meaning systems. Among them are George Kelly's (1955) repertory grid technique; multidimensional scaling (Kruskal, 1964; Messick, 1956; Shepard, 1962; Torgerson, 1952); and points-of-view analysis (Tucker & Messick, 1963). With the use of these powerful multivariate techniques, we might be able to obtain a clearer picture of the organization of time concepts within individuals.

8. Temporal Characteristics and Orientations as Traits and States

Many studies have demonstrated correlations between individual temporal orientations and various personality characteristics. In several of our own studies (Gorman, 1971; Gorman & Wessman, 1974; Thayer, Gorman, Wessman, Schmeidler, & Mannucci, 1975; Wessman, 1973; Wessman et al., 1974), it has been shown that different measures of individual temporal orientations are related to many projective and self-report personality measures. In most, if not all, of these studies we had viewed individual differences in personality orientations as relatively permanent and relatively transituational personality "traits."

For example, one such study (Wessman, 1973) employed factor analysis of responses to a 201-item temporal experience questionnaire from 110 subjects, combined with intensive assessment research on a subsample of 17 subjects to investigate the personality correlates of the factors. Four main temporal experience factors were extracted. An *immediate time pressure* factor, contrasting harassed lack of control versus relaxed mastery and control, was correlated with high emotionality and nervous tension, imaginative fantasy and self-absorption, and sensitivity. A *long-term personal direction* factor, contrasting continuity and steady purpose versus discontinuity and lack of direction, was correlated with happiness and elated mood levels, as well as self-esteem and identity. A *time utilization* factor, contrasting efficient scheduling versus procrastination and inefficiency, was correlated with precision and orderliness, as well as confidence and initiative. A *personal inconsistency* factor, contrasting inconsistency and changeability versus consistency and dependability, was correlated with affective lability and low repression and with impulsiveness.

Three of the temporal experience factors emerging in this study were very close in content and meaning to the factors found in an independent but similarly conceived study of personality and time attitudes by Calabresi and Cohen (1968). We have found these factors to replicate fairly well in subsequent unpublished studies with different groups. Our work, plus research by many others whom we have already cited, makes us feel reasonably certain that it is appropriate to some degree to regard temporal orientations, attitudes, and experiences as persisting, characteristic personality traits. Such general and established personal styles of experiencing and structuring time clearly do appear to have long-term significance and to show meaningful relationships to other psychological characteristics. However, we are not content to conceive of temporal characteristics as essentially fixed and unchanging personal orientations or traits. The dynamic and changing aspects of temporal orientation must be recognized.

In recent years, the traditional concepts of personality traits has fallen into disfavor among many research-oriented psychologists. Several considerations have prompted the adoption of concepts other than traits as useful descriptive personality constructs. For one thing, the ability to predict individual behaviors across situations is often fairly weak. For example, most personality measures do not correlate much above 0.30 with external validating criteria. In addition, measures of the same personality traits do not correlate very highly across differing situations. It is difficult to ask a research subject to answer meaningfully a question such as "I am a nervous person. A. True B. False" without the subject's legitimately asking us to qualify our question further by stating the situation in which the person is "nervous."

Personality researchers and theoreticians such as Raymond Cattell (1965), Walter Mischel (1968), and Charles Spielberger (1966) have suggested that it might be useful to consider the concept of personality *states* to cover the rather transient and situation-specific aspects of behavior and to reserve the term *traits* for more permanent and transituational aspects of behavior. Some notable examples of personality "states" can be seen in moods (Wessman & Ricks, 1966). A person may feel anxious or guilty or sad at a particular time and it would probably not be correct to assume that these feelings are persisting traits. To do so would be to conclude that the person is *always* or *generally* anxious or guilty or sad. It may, in fact, be true that the person is consistently anxious, guilty, or sad, and in this case, we could legitimately speak of personality traits. However, it should be noted that we have no compelling reason to assume personality traits unless we can demonstrate consistency over time and situations.

Although we, like most other investigators, have generally assumed that various temporal orientations are traits, we have not really tested this assumption. It may very well be the case that temporal orientations and experiences are actually states. For example, a person may feel very happy, energetic, and self-confident, and he may tell us that he sees events as flowing rapidly, that he sees the future as promising, and that he spent very little time thinking of his past today. Yesterday, however, the same person might have felt depressed and disappointed, and he might have told us that he saw time as dragging, that he saw the future as threatening and gloomy, and that he dwelled upon many past events. To assign trait labels such as "future-oriented," "past-oriented," or "rapid" versus "slow temporal flow" to this person would be unjustified.

One very powerful method that could be used for examining state and trait effects in time experiences can be found in p-technique factor analysis (Cattell, 1965; Wessman & Ricks, 1966, pp. 71–89; Gorman, 1971; Gorman & Wessman, 1974). In this technique, an individual subject is asked to supply ratings on a preselected set of scales on each day for a period of several days, weeks, or months. For example, a subject is asked to rate himself on elation–depression, energy–fatigue, tranquility–anxiety, etc., each night before he goes to sleep. The subject regularly supplies reports over an extended period and therefore gives us a data matrix in which the rows of the matrix are days and the columns of the matrix are moods, states, or other events.

Several important things can be obtained from the p-technique data matrix. By calculating the mean or median of the person's daily ratings on a given scale, we can obtain an index of the person's overall level on a given state such as anxiety or happiness. By calculating the range or variance of the person's ratings on a given state over a given time period, we can tell how much the person's ratings varied around his own average level. We would be on reasonable grounds if we labeled someone as having the trait of "happiness" if the mean of all of his daily ratings of his states of "happiness" was high and the variance of his daily ratings was sufficiently low. However, a large amount of variability would militate against any simple trait label. By intercorrelating the various ratings of states over a given time period, we can examine how states covary within a single individual over time. For example, suppose we asked the person to rate elation versus depression each day and also asked the person to rate the extent to which he thought about the future and the past. Intercorrelations among the states of elation–depression, "pastness," and "futurity" would tell us how these states varied within the person. For example, the person's data might show a large positive correlation between elation and "futurity" and large negative correlations between elation and "pastness" and "pastness" and "futurity." Such data would indicate his general tendency to think more about the future and less about the past when happy and to do the reverse while depressed. A data reduction technique such as factor analysis or cluster analysis could then be applied to the intercorrelation matrix in order to detect broad patterns of state covariation.

Because p-technique analysis is based on within-individual data rather than on data obtained from a group of individuals, the technique is ideally suited for detecting individual and unique patterns of inter-relationships in meaning systems and psychodynamics. If we obtain p-technique factor analysis matrices from many different people and we find that the same factors tend to emerge from each matrix, we could confidently say that a given state (e.g., elation) generally varies with another state (e.g., futurity).

As far as we know, there have been no studies in which aspects of time experience have been used as variables in p-technique designs. As previously suggested, some useful scales of temporal states as well as traits might be obtained by the combination of some of Osgood's semantic differential dimensions of evaluation, potency, activity, and concreteness with the concepts of past, present, and future. For example, we might build an item or a scale for evaluation of the past (e.g., "Today I feel that things that happened to me in the past were good"); for evaluation of the future (e.g., "Today I believe that my future will be good"); or for concreteness of the past (e.g., "Today I vividly recalled events from my past"). At least 15 temporality scales could be derived by a combination of the Osgood dimensions and temporal zones. Measures of these states could be intercorrelated with each other and with measures of other states, such as moods and physiological indicators.

Although we do not as yet have proof that temporal concepts considered as states are more valuable in personality and motivation studies than temporal trait conceptualizations, we do have some rather promising leads. In mood and personality studies, Wessman and Ricks (1966, pp. 33–54) found that, in a Q-sort task, people were more likely to consider their present "actual" self-concept ratings as closer to their "ideal" self-concept ratings when they were elated than when they were depressed. As the ideal self-concept rating task had an explicit future orientation (". . . a picture of the sort of person you hope to become or fancied yourself to be . . . what you would wish to be" [Wessman and Ricks, 1966, p. 40]), we have some indication that some temporally-oriented concepts vary with mood states. Similarly, Harrow and Ferrante (1969) gave Rotter's *Internal versus External Locus of Control* questionnaire (Rotter, 1966) to manic–depressive patients and found that when these patients were depressed, they tended to endorse beliefs that indicated that the major successes and failures in their lives were due to luck, chance, fate, and the influence of other people. However, when manic, the patients endorsed beliefs attributing rewards to skill and personal responsibility. As locus of control is significantly correlated with temporal concepts (Thayer *et al.*, 1975) and locus of control may, at least partially, be a state variable, we again have some indirect evidence of the feasibility of employing temporal state measures in further studies.

9. The Utility of Time Concepts

In previous sections of this chapter, we have discussed the development of time concepts, their representation, and the possibility that they might be state rather than trait variables. However, we have only briefly alluded to the functions of temporal concepts for the person. Time concepts are undoubtedly constructed and selected for important purposes and adaptive ends. We believe that time concepts permit the person to erect a meaningful and coherent framework for living and acting from the welter of happenings that impinge upon him and in which he participates. We have stressed the importance of examining differences in temporal concepts between individuals and within individuals. However, few studies seem to address themselves to the issues of *why* and for *what purpose* time concepts are formed.

In many studies, temporal characteristics are found to differ between groups, but we are seldom given any explanation of the reasons for these differences. Apparent differences in time concepts might be based on very similar motives, yet sometimes apparent identity in time concepts between groups of individuals might, in fact, be serving quite different motives and purposes. For example, Leonard Doob (1971, pp. 16–17) gave a cogent illustration of the necessity for careful examination of the reasons behind temporal behaviors when he discussed the meaning of delayed responses in delay of gratification studies. Doob asked a group of African subjects whether they would prefer to receive a small amount of money immediately or a larger amount of money one year later. Most subjects tended to choose the small immediate reward. At first glance, we might be led to conclude that these subjects displayed foreshortened time perspectives and little delay of gratification. For some subjects, this might have been true. However, some of the subjects reported that they would take the small immediate reward and invest it so that they would receive a return larger than the long-term reward offered by the experimenter! It can be seen here that a researcher who examines carefully the utility of the time perspective for his subjects gains much more than if he merely classifies their behavior in terms of simple preference for immediate versus long-term rewards.

In this volume, Kastenbaum and Cottle have clearly shown that the framing of personal and social events in temporal perspectives are strongly guided by the needs and desires of the persons involved. For example, Cottle described two boys who could both be considered "future-oriented." However, the poor boy saw his future as unrewarding and frustrating and therefore felt that he would be wasting his time making future plans. A middle-class suburban boy, however, saw his future as the fulfillment of promises and indulged in fantasies of the future that the poor boy would have found too painful to consider. Kastenbaum gave examples of the very complex sets of goals and purposes that older adults may demonstrate in maintaining temporal perspectives and showed that one cannot simply classify an older person as past-, future-, or present-oriented without exploring the reasons that the person maintains such a stance. In general, then, individual perspectives on time are constructed as guideposts and markers for the person's own meaning system and are formed to serve one's sometimes unique pattern of needs.

10. Time and Psychopathology

Perhaps the most critical need for better explorations of temporal concepts can be seen in psychopathology research. Confronted with all of the complexities of human experience, the clinician is often pressed to apply the as-yet-incomplete methods and theories of personality change to disorders whose etiologies are often only vaguely understood. As time experiences are vital aspects of human existence, it seems natural to explore the roles of temporal experiences in the clinical disorders. By examining subjective time in pathological conditions, we can gain a fuller understanding of both normal and abnormal states.

While there have been a large number of articles on time experiences in specific mental disorders, it appears to us that many of the articles have not supplied the reasons for the often altered and bizarre constructions of time that tend to appear in the various disorders. We feel that the lack of functional explanations of time experiences in pathology is not peculiar to time research alone but can be found in many areas of clinical research. We discuss here some of the salient findings of time research in some of the disorders, and where appropriate, we offer some conjectures on the role of time experience in the disorders.

10.1. Schizophrenia

One of the persistent problems in psychology is the problem of explaining schizophrenia. Studies of time estimation in schizophrenia have shown that schizophrenic patients seem to be generally inaccurate in their estimates of both long and short time intervals and tend to display great variability in their estimates of time intervals (Dobson, 1954; Johnson & Petzel, 1971; Wallace & Rabin, 1960). Schizophrenics also seem to be disoriented in their time concepts and frequently confuse past, present, and future events (De La Garza & Worchel, 1956; Schlosberg, 1969; Wallace, 1956) more than matched control subjects. Schizophrenics are also likely to give more remote temporal associations than normal subjects (Orme, 1969). One might question, however, whether these findings are unique to *temporal* experiences in schizophrenia. There is substantial evidence to show that schizophrenics display such cognitive deficits as overinclusive thinking, high distractability, inability to filter out extraneous information, poor perceptual constancy, and deficiencies in being able to maintain perceptual sets for any length of time (McGhie, 1969). All of the temporal alterations in schizophrenia seem to be related to the broader cognitive problems suffered by schizophrenics and do not seem to be limited to the experience of time alone.

10.2. Neurotic Disorders

In neurotic patients, we do not see as many systematic disturbances in the estimation of short time intervals as we do in schizophrenics, but researchers have found some. Several studies have indicated that anxious normal and neurotic subjects are likely to estimate the passage of clock time as faster than the actual clock time (Falk & Bindra, 1954; Siegman, 1962; Whyman & Moos, 1967). However, Hare (1963) found that when anxious subjects were asked to count off elapsed time, they were actually more accurate than less anxious subjects. Unlike Hare, Whyman and Moos (1967) found that anxious subjects displayed greater inaccuracy in time estimation than less anxious subjects and also showed greater variability in their judgments. Because the ways of assessing anxiety and the methods used to obtain time estimates differed from study to study, it is difficult to say why the various results were obtained. It may be that the relationship between accuracy in time estimation and anxiety follows the "inverted-U" function often found in many motivation and performance studies. If so, both low anxiety and high anxiety may be associated with inaccurate estimation, while moderate levels of anxiety might lead to greater accuracy. It may be that under different levels of anxiety arousal, different amounts and kinds of cues are employed in the forming of time judgments. It may also be that when people are anxious, they are so preoccupied with more pressing needs and tasks that they are inattentive to cues that might supply better time estimates.

Studies of the relationships between time perspective and anxiety in normal and neurotic people tend to indicate that in comparison to less anxious subjects, anxious subjects are more likely to be preoccupied with the present than with the past and the future (Epley & Ricks,

1963; Gorman, 1971; Gorman & Wessman, 1974; Krauss & Ruiz, 1967; Rychlak, 1972; Wessman, 1973; Wessman & Ricks, 1966; Wessman *et al.,* 1974). However, Meerloo (1970) and Lipman (1957) have spoken about neurotics who are both highly anxious and highly concerned with future events.

Several possible processes may mediate the relationship among time perspectives and anxiety. One possibility may lie in the meaning of anxiety to the subjects. For example, Krauss (1967) has defined anxiety as the "fear of a future event," and several researchers (Calabresi & Cohen, 1968; Cottle, 1969; Wessman, 1973) have shown that psychometric scales of "time anxiety" or "time pressure" that contain items relevant to fear of future events and feelings of lack of control correlate highly with other measures of anxiety. It may be that some neurotics, with strong anticipations of danger and punishment in the future, may tend to repress or suppress thoughts of the future, while others may tend to attempt some mastery over possible threats by ruminating about the future.

Very little work has been done on the temporal aspects of subcategories of neurotic disorders. Among the more salient neurotic disorder subtypes are the phobic, anxiety state, obsessive–compulsive, and hysterical neuroses. Although there has been much clinical literature, little empirical research has been done to examine directly the time experiences of obsessive–compulsives. Several indirect lines of evidence point to the notion that obsessive–compulsive people are highly concerned with a need to master time by maintaining tightly rigid schedules, which seem to reduce the dread of empty time and unexpected, uncontrolled events (Fenichel, 1945; Gorman & Katz, 1971; Pettit, 1969; Meerloo, 1966). By rigidly forming and then maintaining schedules with technical precision, the obsessive–compulsive seems to be following the many "shoulds" and "oughts" that he has constructed in order to avoid loss of control, with its resulting shame and uncertainty. It is highly unlikely that the obsessive–compulsive would be deficient in extending his thoughts into the future, and several astute clinical observers, especially Arieti (1967) and Shapiro (1965), have pointed to the tense deliberation and extensive plans typically made by the obsessive–compulsive. However, it seems that the obsessive–compulsive person becomes so bogged down by the minutiae and the sheer number of his plans that he has a very small chance of actually executing any of them! It may be that the very act of planning may be more important for the obsessive than the actual outcome of the plan.

Observations of hysterical patients as early as Freud and Breuer's (1966) *Studies on Hysteria* have indicated that hysterics act out their impulses in both direct and symbolic forms in the present while repressing past events. Shapiro (1965, p. 111) suggested that the cognitive style of the hysteric is global, relatively diffuse, and highly impressionistic. Shapiro noted that the hysteric is "struck" by events and displays none of the cautiousness and delay seen in other neurotic patients, especially obsessive–compulsives. With the emphasis on immediately present events and the strong use of repression, we should expect the hysterical patient to be highly present-oriented and hardly aware of the future and the past at all. However, dissociative forms of hysteria may present a problem, as the person may develop a different temporal orientation for each personality state. In the dissociative states, we would probably see a fragmentation and lack of integration of temporal "zones," although this effect remains to be validated.

10.3. Affective Disorders

A considerable amount of literature has been written about the experience of time in the affective disorders. Many clinicians have agreed with Minkowski (1958) and Straus (1947) that depressed people tend to see the present as stagnant, nonflowing, and rigid; the past as remote and detached from present concerns but still, paradoxically, important; and the future as hopeless, unrealizable, and threatening. Dilling and Rabin (1967) found that in comparison to schizophrenics and "normal" control subjects, depressives display significantly less future extension. Wessman and Ricks (1966) found that among normal college students, happy men were more likely to be future-oriented than less happy men. Using several methods, several studies indicated that depressive patients may experience personal time as dragging when compared to clock time (Dilling & Rabin, 1967; Lhamon, Goldstone, & Goldfarb, 1965; Mezey & Cohen, 1961). However, the time estimation studies are not highly consistent in their findings and employed many different methods and somewhat different time intervals.

Perhaps the decline of future extension in depressives and the occasional clinging to past concerns can be explained in several ways. Bibring (1953), Beck (1967), and Seligman (1975) have each considered feelings of helplessness, fears of inferiority, and lowered self-esteem to be the major signs of depression. Since self-esteem and related feelings of competence are often based on past achievement, present states, and future aspirations, the depressive person may attempt to retrace the steps leading from the past to his present feelings. Since the present situation would provide the background for future expectations, the person has several choices at his disposal: (1) he can expect the future to be as disappointing as the present and can avoid thoughts of the future; (2) he can deny present disappointments and past failures and "leap" into a future into which great (sometimes manic and grandiose!) hopes and confidence are projected; (3) he may form a delusional set of constructions, as in suicidal ideations, in which he may project magical wishes into an afterlife that promises relief from present woes; (4) he may reconstruct memories of a real or a fantasized happy past; and (5) he may realistically cope with the present situation, maximizing opportunities for happiness and investing in future successes.

Kastenbaum (1965) asked subjects to supply endings for incomplete stories that varied in terms of affective content; the use of past, present, and future tenses; and individual versus group content. Kastenbaum found that although most story endings extended into the future, the introduction of both pleasant and unpleasant affective content increased the likelihood of past endings but had no effect on the probability of future endings. Kastenbaum provided an interesting interpretation for his results. He observed that the "easiest" way for a person to complete a story can be found in the person's continuation of the narrative in a future direction (e.g., "and so it goes"). However, the more difficult way of ending a story involves reversing the flow of the story by supplying past endings. In light of Kastenbaum's observations, it may be that depressed patients are attempting to cope actively with negative affective material by taking the more difficult route of delving into their pasts.

The subjective experiences of the rate of time flow in depressive and manic disorders present an interpretive problem that may possibly be explained in several ways. It may be that when people are intensely aroused by pleasant or unpleasant feelings, they are unlikely to attend to cues that would lead to more accurate correspondence to clock time. It may be that under arousal states, the person may be attending to the acceleration of the activity of his autonomic nervous system and may be translating the judgments of the speed of his physiological reactions (see Voyat, this volume) into judgments of "time." It may also be the case, as Ornstein (1969) suggested, that under arousal, the physiological structures responsible for the cognitions that we call "time" are altered in such a way as to produce distortions in judgments of time as well as many in other cognitions.

10.4. Personality and Character Disorders

Character disorders are usually described in terms of failures of the socialization process. For various reasons, the person with disorders in this group has either failed to learn the traditional and acceptable controls, customs, and roles of the society or has been socialized into the ways of a deviant subculture. Because many temporal concepts are socially transmitted and play important roles in self-control and in interactions with others, we should predict that deviant socialization patterns and deviant temporal conceptualizations should be related. Several lines of evidence lead to the confirmation of this notion. Barndt and Johnson (1955); Davids, Kidder, and Reich (1962); Stein, Sarbin, and Kulik (1968); and Seigman (1961) all found that delinquent boys displayed significantly less future time extension than nondelinquents. Ricks, Umberger, and Mack (1964) reported that delinquents who had been successfully treated in psychotherapy tended to produce TAT stories with longer future and past time perspective than unsuccessfully treated delinquents. Hare (1970, 1974) found that psychopaths show very little anticipatory electrodermal activity prior to the onset of a painful electric shock. Hare's finding suggests a possible physiological difference among psychopaths and nonpsychopaths, which may explain some of the reasons that psychopaths seem to be fairly unperturbed in the face of impending threat. However, Hare's findings must be tempered by the fact that psychopaths may actually show considerably more cardiovascular activity prior to an electric shock than controls.

Shapiro (1965) discussed impulsive character styles and pointed out that the highly impulsive person may be quite effective in carrying out short-range plans but quite deficient in carrying out long-range plans. These people may experience sporadic whims that are poorly integrated into long-range plans and ongoing actions. Shapiro noted that it was rare to find an impulsive character with a strong set of purposeful goals beyond his own immediate needs. Although not originally designed to test Shapiro's notion of impulsive character styles, several studies point to the notion that delay of gratification, impulse control, and time perspective have a common cognitive substrate. In several studies, Mischel and his colleagues (Mischel, 1968; Mischel & Metzner, 1962) gave children the choice of receiving a small but immediate reward or a larger but delayed reward for successful task performance. It was found that younger children were more likely to choose the immediate reward than the delayed reward, while older children were more likely to choose the delayed reward. Rozek, Wessman, and Gorman (1977) repeated Mischel's procedure in groups of 4–5, 6–7, and 9-year-old girls and found that there was a strong set of relationships between the child's age, the child's preference for delayed reward, and the child's tendency to complete stories with extended future time perspectives. Children who chose delayed rewards also seemed to display a greater understanding of temporal concepts, more accurate production estimates of a one-minute duration, and greater conservation of number. A strong relationship between age and delay of gratification was found, so that at present, we cannot definitely state that delay of gratification is due purely to cognitive developmental factors. However, it is clear that for whatever reason, cognitive development and delay of gratification are strongly related.

10.5. Organic Brain Syndromes and Mental Retardation

As time experiences are cognitive products, it is not surprising that injuries to the parts of the nervous system that are involved in perceptual, judgmental, attentional, and memory processes tend to produce alterations in temporal experiences. Several studies (e.g., Benton, Van Allen, & Fogel, 1964; Coheen, 1950; Davidson, 1941) have shown that damage to such diverse areas as the parietal, the temporal, and the frontal lobes produces temporal distortions and inaccuracy. Not surprisingly, the more extensive and the more diffuse the organic damage was, the more the likelihood of temporal disorientation. To date, no specific brain region has been shown to be uniquely responsible for temporal disorientation, and it is unlikely, given the nature of brain structures, that any such area will be found. Instead, it will probably be the case that future brain studies will find systems of brain structures that produce the cognitive deficits underlying temporal distortions.

As with brain damage, we find that in mental retardation there are disorientations in time, confusion in temporal conceptualizations, and inaccuracy in time judgments. As can be expected from Piaget's studies of the development of time concepts (Piaget, 1966, 1970; Voyat, this volume; Wessman, this volume), retardates display shorter future time extension (Roos & Albers, 1965); less developed concepts of simultaneity, temporal order, and equality of synchronized intervals (Lovell & Slater, 1960; Montroy, McManis, & Bell, 1971); and are less accurate in time estimation (McNutt & Melvin, 1968) than nonretarded subjects of the same chronological ages. The Lovell and Slater (1960) and Montroy *et al.* (1971) studies indicated that consistent with Piagetian theory, retardates develop time concepts in the same order as normal subjects but do so at a slower rate and at later chronological ages.

In general, studies of temporal experience in psychological disorders can provide important contributions to our understanding of time. However, we must continually ask how particular temporal experiences play a role in the various disorders. Merely to catalog findings without some theoretical rationale will result in a warehouse of isolated data that probably will not advance our knowledge of time.

Psychopathology researchers are constantly confronted with disorders that are often discussed in terms of social values. As we shall see in the next section, the value orientations of the researcher can have a profound influence on his tactics of investigating and interpreting temporal research.

11. Research Values and Time Research

Throughout this chapter, we have spoken of time in terms of the processes of construction of cognitive schemata and systems of personal meaning, and we have attempted to stress the relativity of the meanings of "time" for individuals and societies. We have suggested that time concepts are rooted in personal and social networks of meanings and values. One pitfall, however, of conducting research on the underlying motives behind time construction can be found in the tendency to inject our own values into time research.

We know that the social scientist cannot conduct research in a totally dispassionate and objective way. Although it would be ideally desirable to conduct value-free research, the very acts of choosing topics of investigation, deriving hypotheses, and choosing research methodologies are largely determined by the researcher's own value system and personal cosmology. Values and viewpoints are products of the socialization forces to which the researcher has been exposed. Among these viewpoints are many implicit beliefs concerning the relative "goodness" or "badness" of one mode of experience as compared to another.

We find that personality theories and research often have an implicit and sometimes explicit "good guys versus bad guys" emphasis in which one mode of behavior is regarded more favorably than some other type of behavior. As time concepts seem to be highly loaded with evaluative components, it seems difficult to avoid prescriptions for the "good life" in one form or another of time structuring. For example, in our society there seems to be a generally negative evaluation of past-related concepts, with words like *repressive, backward, regressive, recalcitrant, refractory,* and *regret,* clearly indicating the undesirability of past events or orientations. Future terms, though, are generally seen as positive, with words like *forward-looking, progressive,* and *foresight* having highly favorable connotations. Such evaluative statements often emerge in discussions of individual differences in past, present, and future time perspectives. However, we might ask if we can go beyond these evaluations of time and see some positive functions of each time perspective.

In an article entitled "A Funny Thing Happened on the Way to the Future," Warren Bennis (1970) said:

> For me the 'future' is a portmanteau word. It embraces several notions. It is an exercise of the imagination which allows us to compete with and try to outwit future events. Controlling the anticipated future is, in addition, a social invention that legitimizes the process of forward-planning. . . . Most importantly, the future is a conscious dream, a set of imaginative hypotheses groping toward whatever vivid utopias lie at the heart of our consciousness. (p. 431)

Unlike others, who have greeted the "future" simply with optimism or pessimism, Bennis was able to show that the notion of the future varies considerably and that prescriptions for future behaviors serve very different values. For example, Bennis spoke of various value orientations such as : (1) a *militant orientation* that seeks to meet the future by destroying the *status quo*; (2) an *apocalyptic orientation* in which the person takes a pessimistic stance toward any improvement and prophesies impending "gloom and doom"; (3) a *regressive orientation* in which the present is seen as decaying and there is a strong emphasis on nostalgia; (4) a *retreatist orientation* in which the person meets changing situations by displaying apathy and by withdrawing in favor of inner experiences; (5) a *historical orientation* in which the person seeks to demonstrate that the "good old days" were either far better or far worse than the present; (6) a *technocratic orientation* expressed by the valuing of technical progress; and (7) a *liberal-democratic orientation* in which the future is met with optimism concerning the perfectability of Mankind through social reforms and the logical, scientific conquest of unreason.

A striking feature of Bennis's typology lies in the fact that *all* of these future ideologies confront individuals in everyday life and most of them have some positive virtues as well as some obvious faults. For example, in our current political and economic turmoil the individual is regularly admonished to "work for reform" (militant, liberal-democratic); worry about the depletion of natural resources (apocalyptic, technocratic); "remember our national heritage" (regressive, historical); believe that "progress is our most important product" (technocratic); "turn on, tune in, drop out" (retreatist); and to "dream the impossible dream" of social perfectability (liberal-democratic). To obey all of these value injunctions at once, one would have to live an inconsistent and fragmented existence. However, we often see research interpretation and social analysis couched along one or the other of these value dimensions, in which a particular stance toward the future is considered to be the "good" one.

We agree with Bennis that the concept of "the future" and of time perspective, in general, is a "set of imaginative hypotheses." As personality researchers, our task should be one of critically examining how social values are expressed in time perspectives while attempting to maintain our own objectivity by avoiding an overly simple "good guys versus bad guys" stance ourselves.

12. The Past, Present, and Future of Time Research

The tenor of this chapter may appear critical and even somewhat pessimistic. But this is not entirely the case. We hope that our criticisms are a sign of our "growing pains" and the harbingers of better things to

come. With this hope in mind, let us look at the growth of ideas about
time. One way in which we might be able to examine the shift of
paradigms in time research can be found in adopting a genetic epistemological viewpoint about the growth of ideas in philosophy and the
sciences. In several of his writings, Piaget (1970, 1971) has expressed
the notion that developments in philosophy, science, and technology
often parallel the developments of thought and logic in the individual
child. Thus, "thought" has its own infancy and adulthood and we can
see relatively more or less developed modes of thought as a science or
field advances.

In the early sensorimotor stages of cognitive development, concepts
are crude and are concretely represented as palpable, easily seen, and
easily manipulated "things." For example, the infant of 3 months of age
"knows" a face only when it is placed directly in front of him. A similar
development seems to have occurred in conceptualizations of time, in
which earlier researchers were concerned with "viewing" time as a
sensory object and then searched for the sensory "organs" capable of
processing a "time sense."

As the child enters the preoperational stage of development, he is
able to handle objects concretely. He can entertain the possibility that
objects that are not immediately present can still exist independently of
his seeing them, and he can actively attach labels to objects and
represent them in language. Here too, we see a parallel development in
time research, in which researchers attempted to catalog the different
metaphors and symbolizations of time in order to understand time in
much the same way that the preschooler discovers the names of things
and attempts to construct reality by labeling objects.

At around 6–8 years of age, the child enters the concrete operational stage of development, in which he can perform simple logical
problems, reason in analogies, and conceptualize relationships. While
the 5-year-old cannot understand that a piece of clay can still have the
same mass regardless of how it is stretched or compressed, the 8-year-
old, who has reached the concrete operational stage, has little difficulty
with this task and tells us that the clay's mass is the same but that its
appearance has changed. The 8-year-old, however, would be puzzled by
abstract and intangible tasks, such as solving an algebraic equation or
defining an abstract term such as *democracy*. Time research also seems
to have gone through its own concrete operational stage. In time
research, this stage of development was one in which researchers were

concerned with correlating subjective time perspectives with more "objective" time indicators, with demographic and with biographical variables. Here, the research focused upon the task of finding simple equations that would reconcile subjective impressions of time with more tangible indicators of time.

Finally, Piaget's most advanced stage, the formal operational stage, enables the person to handle highly abstract concepts that have few, if any, tangible referents. For example, the person in this stage of development can handle algebraic problems, symbolic logic, and verbal abstractions without the necessity of concrete and manipulable symbols. Although some rudiments of the ability to handle a multiplicity of interacting variables from a variety of viewpoints can be found in earlier stages, the multivariate nature of thought reaches its fullest development in this stage. Throughout this chapter, we have been arguing that perhaps we have advanced in our thinking about time to the point where we can now entertain more formal operational modes of thought. Hopefully, we can accept a relativistic viewpoint in which we can examine how temporal experiences are components of larger networks of ongoing processes. Such a stance is not easy to adopt because it means that we have to rely less upon the concepts of "objective" time and the use of psychophysical methodologies. We cannot be overly concerned with a "time sense" *per se* but must be concerned with the motives for time reckoning, the functions of different temporal perspectives for individuals and societies, and the ways in which temporal variables interact with other variables. We must look at individual differences in time conceptualizations as states as well as more enduring traits. We must view time not as a universal "given" but in relation to the individual or culture, which constructs time for its own purposes.

Let us go back to St. Augustine's question, "What is time?" and, still appreciating its importance, open the path for our studies by asking a corollary question: "How then, do we use time?"

References

Abraham, K. *Selected papers.* London: Hogarth, 1927.

Arieti, S. *The intrapsychic self: Feeling and cognition in health and mental illness.* New York: Basic Books, 1967.

Augustine, St. *St. Augustine's confessions.* W. Watts (Trans.). Cambridge: Harvard, 1912.

Barndt, R. J., & Johnson, D. M. Time orientation of delinquents. *Journal of Abnormal and Social Psychology,* 1955, *51,* 343–345.

Beck, A. T. *Depression: Clinical, experimental, and theoretical aspects.* New York: Harper and Row, 1967.

Bennis, W. G. A funny thing happened on the way to the future. *In* F. F. Korten, S. W. Cook, & J. I. Lacey (Eds.), *Psychology and the problems of society.* Washington, D.C.: American Psychological Association, 1970.

Benton, A. L., Van Allen, M. W., & Fogel, M. L. Temporal orientation in cerebral disease. *Journal of Nervous and Mental Disease,* 1964, *139,* 110–119.

Bergler, E., & Roheim, G. Psychology of time perception. *Psychoanalytic Quarterly,* 1946, *15,* 190–206.

Bibring, E. The mechanism of depression. *In* P. Greenacre (Ed.), *Affective disorders: Psychoanalytic contributions to their study.* New York: International Universities Press, 1953.

Bonaparte, M. Time and the unconscious. *International Journal of Psychoanalysis,* 1940, *21,* 427–468.

Brandon, S. G. F. *History, time, and deity.* Manchester: Manchester University Press, 1965.

Buss, A. R. An extension of developmental models that separate ontogenetic changes and cohort differences. *Psychological Bulletin,* 1973, *80,* 466–480.

Buss, A. R. Generational analysis: Description, explanation, and theory. *Journal of Social Issues,* 1974, *30,* 55–72.

Calabresi, R., & Cohen, J. Personality and time attitudes. *Journal of Abnormal Psychology,* 1968, *73,* 431–439.

Cattell, R. B. *The scientific analysis of personality.* Baltimore: Penguin, 1965.

Coheen, J. L. Disturbances in time discrimination in organic brain disease. *Journal of Nervous and Mental Disease,* 1950, *112,* 121–129.

Cohen, J., Hansel, C. E. M., & Sylvester, J. D. An experimental study of comparative judgements of time. *British Journal of Psychology,* 1954, *45,* 108–114.

Cottle, T. J. The circles test: An investigation of temporal relatedness and dominance. *Journal of Projective Techniques and Personality Assessment,* 1967, *31,* 58–71.

Cottle, T. J. Temporal correlates of the achievement value and manifest anxiety. *Journal of Consulting and Clinical Psychology,* 1969, *33,* 541–550.

Cottle, T. J., & Klineberg, S. L. *The present of things future.* New York: Free Press, 1974.

Davids, A., Kidder, C., & Reich, M. Time orientation in male and female juvenile delinquents. *Journal of Abnormal and Social Psychology,* 1962, *64,* 237–240.

Davidson, C. M. A syndrome of time agnosia. *Journal of Nervous and Mental Disease,* 1941, *94,* 336–337.

De La Garza, C. O., & Worchel, P. Time and space orientation in schizophrenia. *Journal of Abnormal and Social Psychology,* 1956, *52,* 191–194.

Dilling, C. A., & Rabin, A. I. Temporal experience in depressive states and schizophrenia. *Journal of Consulting Psychology,* 1967, *31,* 604–608.

Dobson, W. R. An investigation of various factors involved in time perception as manifested by different nosological groups. *Journal of General Psychology,* 1954, *50,* 277–298.

Doob, L. *Patterning of time.* New Haven, Conn.: Yale University Press, 1971.

Ellenberger, H. F. A clinical introduction to psychiatric phenomenology and existential analysis. *In* R. May, E. Angel, & H. F. Ellenberger (Eds.), *Existence: A new dimension in psychiatry and psychology.* New York: Basic Books, 1958.

Epley, D., & Ricks, D. F. Foresight and hindsight in the TAT. *Journal of Projective Techniques,* 1963, *27,* 51–59.

Erikson, E. H. *Childhood and society* (2nd ed.). New York: Norton, 1963.

Erikson, E. H. *Identity, youth, and crisis.* New York: Norton, 1968.

Falk, J. L., & Bindra, D. Judgement of time as a function of serial position and stress. *Journal of Experimental Psychology,* 1954, *47,* 279–282.

Fenichel, O. *The psychoanalytic theory of the neuroses.* New York: Norton, 1945.

Fraisse, P. *The psychology of time.* New York: Harper and Row, 1963.

Fraser, J. T. *Of time, passion, and knowledge.* New York: Braziller, 1975.

Freud, S. *The ego and the id.* J. Riviere (Trans.). New York: Norton, 1960. (Originally published, 1923.)

Freud, S. *The interpretation of dreams.* J. Strachey, (Ed. and Trans.). New York: Avon, 1967. (Originally published, 1900.)

Freud, S., & Breuer, J. *Studies on hysteria.* J. Strachey & A. Strachey (Eds. and Trans.). New York: Avon, 1966. (Originally published, 1895.)

Fromm, E. *Man for himself.* New York: Holt, Rinehart and Winston, 1947.

Gilliland, A. R., Hofeld, J., & Eckstrand, G. Studies in time perception. *Psychological Bulletin,* 1946, *43,* 162–176.

Goldstein, K. *The organism.* New York: American Book Company, 1939.

Gorman, B. S. A multivariate study of the relationship of cognitive control and cognitive style principles to reported daily mood experiences. Doctoral dissertation, City University of New York. Ann Arbor, Mich.: University Microfilms, 1971, No. 72-5071.

Gorman, B. S., & Katz, B. Temporal orientation and anality. *Proceedings of the 79th Annual Convention, APA,* 1971, 367–368.

Gorman, B. S., & Wessman, A. E. The relationship of cognitive styles and moods. *Journal of Clinical Psychology,* 1974, *30,* 18–25.

Gorman, B. S., Wessman, A. E., Schmeidler, G. R., Thayer, S., & Mannucci, E. Linear representation of temporal location and Stevens' law. *Memory and Cognition,* 1973, *1,* 169–171.

Hare, R. D. Anxiety, temporal estimation, and rate of counting. *Perceptual and Motor Skills,* 1963, *16,* 441–444.

Hare, R. D. *Psychopathy: Theory and research.* New York: Wiley, 1970.

Hare, R. D. Psychopathy. *In* P. Venables & N. Christie (Eds.), *Research in psychopathology.* New York: Wiley, 1974.

Harrow, M., & Ferrante, A. Locus of control in psychiatric patients. *Journal of Consulting and Clinical Psychology,* 1969, *33,* 582–589.

James, W. *The principles of psychology.* Vols. 1 and 2. New York: Holt, 1890.

Jenner, F. A., Goodwin, J. C., Sheridan, M., Tauber, I. U., & Lobban, M. C. The effects of altered time regime on biological rhythms in a 48-hour periodic psychosis. *British Journal of Psychiatry,* 1968, *111,* 215–224.

Johnson, J. E., & Petzel, T. P. Temporal orientation and time estimation in chronic schizophrenics. *Journal of Clinical Psychology,* 1971, *27,* 194–196.

Kastenbaum, R. The direction of time perspective: The influence of affective set. *Journal of General Psychology,* 1965, *73,* 189–201.

Kelly, G. *The psychology of personal constructs.* New York: Norton, 1955.

Knapp, R. H., & Garbutt, J. T. Time imagery and the achievement motive. *Journal of Personality,* 1958, *26,* 426–434.

Krauss, H. H. Anxiety: The dread of a future event. *Journal of Individual Psychology,* 1967, *23,* 88–93.

Krauss, H. H., & Ruiz, R. A. Anxiety and temporal perspective. *Journal of Clinical Psychology,* 1967, *23,* 340–342.

Kruskal, J. B. Nonmetric multidimensional scaling: A numerical method. *Psychometrika,* 1964, *29,* 118–129.

Langer, S. K. *Philosophy in a new key: A study of the symbolism of reason, rite, and art.* Cambridge, Mass.: Harvard University Press, 1961.

Lashley, K. *Brain mechanisms and intelligence: A quantitative study of injuries to the brain.* Chicago: University of Chicago Press, 1929.

Lhamon, W., Goldstone, S., & Goldfarb, J. The psychopathology of time judgement. *In* P. Hoch & J. Zubin (Eds.), *The psychopathology of perception.* New York: Grune and Stratton, 1965.

Lipman, R. S. Some relationships between manifest anxiety, "defensiveness," and future time perspective. Doctoral dissertation, University of Connecticut, 1957.

Loehlin, J. C. The influence of different activities on the apparent length of time. *Psychological Monographs,* 1959, *73*(4) (Whole No. 474).

Lovell, K., & Slater, A. The growth of the concept of time: A comparative study. *Journal of Child Psychology and Psychiatry,* 1960, *1,* 179.

Luce, G. G. *Body time.* New York: Pantheon, 1971.

Mach, E. *The analysis of sensations and the relation of the physical to the psychical.* New York: Dover, 1959. (Originally published, 1865.)

McGhie, A. *Pathology of attention.* Baltimore: Penguin, 1969.

McNutt, T. H., & Melvin, K. B. Time estimation in normal and retarded subjects. *American Journal of Mental Deficiency,* 1968, *72,* 582–589.

Meerloo, J. A. M. The time sense in psychiatry. *In* J. T. Fraser (Ed.), *The voices of time.* New York: Braziller, 1966.

Meerloo, J. A. M. *Along the fourth dimension: Man's sense of time and history.* New York: John Day, 1970.

Merleau-Ponty, M. *The phenomenology of perception.* London: Routledge & Keegan Paul, 1962.

Messick, S. J. Some recent theoretical developments in multidimensional scaling. *Educational and Psychological Measurement,* 1956, *16,* 82–100.

Mezey, A. G., & Cohen, S. T. The effect of depressive illness on time judgement and time experience. *Journal of Neurology, Neurosurgery, and Psychiatry,* 1961, *24,* 269.

Minkowski, E. Findings in a case of schizophrenic depressin. *In* R. May, E. Angel, & H. F. Ellenberger (Eds.), *Existence: A new dimension in psychiatry and psychology.* New York: Basic Books, 1958.

Mischel, W. *Personality and assessment.* New York: Wiley, 1968.

Mischel, W., & Metzner, R. Preference for delayed reward as a function of age, intelligence, and length of delay interval. *Journal of Abnormal and Social Psychology*, 1962, *64*, 425–431.

Montroy, P., McManis, D., and Bell, D. Development of time concepts in normal and retarded children. *Psychological Reports*, 1971, *28*, 895–902.

Nichols, H. The psychology of time. *American Journal of Psychology*, 1890, *3*, 453–529.

Orme, J. The time location of associations in schizophrenia. *Journal of Clinical Psychology*, 1969, *25*, 260–261.

Ornstein, R. *On the experience of time.* Baltimore: Penguin, 1969.

Osgood, C. E. Explorations in semantic space: A personal diary. *Journal of Social Issues*, 1971, *27*, 5–64.

Osgood, C. E., Suci, G. J., & Tannenbaum, P. H. *The measurement of meaning.* Urbana, Ill.: The University of Illinois Press, 1957.

Pettit, T. F. Anality and time. *Journal of Consulting and Clinical Psychology*, 1969, *33*, 170–174.

Piaget, J. *Le développement de la notion du temps chez l'enfant.* Paris: Presses Universitaires de France, 1946.

Piaget, J. The development of time concepts in the child. *In* P. Hoch & J. Zubin (Eds.), *Psychopathology of childhood.* New York: Grune and Stratton, 1955.

Piaget, J. Time perception in children. *In* J. T. Fraser (Ed.), *The voices of time.* New York: Braziller, 1966.

Piaget, J. *Genetic epistemology.* New York: Columbia University Press, 1970.

Piaget, J. *Psychology and epistemology: Towards a theory of knowledge.* New York: Viking, 1971.

Pittendrigh, C. S. On temporal organization in living things. *In* H. Yaker, H. Osmond, & F. Cheek (Eds.), *The future of time: Man's temporal environment.* Garden City, N.Y.: Doubleday, 1971.

Ricks, D. F., Umberger, C., & Mack, R. A measure of increased temporal perspective in successfully treated adolescent delinquent boys. *Journal of Abnormal and Social Psychology*, 1964, *69*, 685–689.

Riegel, K. F. History as a nomothetic science: Some generalizations from theories and research in developmental psychology. *Journal of Social Issues*, 1969, *25*, 99–128.

Riegel, K. F. Time and change in the development of the individual and society. *In* H. Reese (Ed.), *Advances in child development and behavior.* (Vol. 7). New York: Academic Press, 1972.

Roos, P., & Albers, R. J. Performance of retardates and normals on a measure of temporal orientation. *American Journal of Mental Deficiency*, 1965, *69*, 835–838.

Rotter, J. B. Generalized expectancies of interval vs. external control of reinforcement. *Psychological Monographs*, 1966, *80*(1) (Whole No. 609).

Rozek, F., Wessman, A. E., & Gorman, B. S. Temporal span and delay of gratification as a function of age and cognitive development. *Journal of Genetic Psychology*, 1977 (in press).

Rychlak, J. F. Manifest anxiety as reflecting committment to the psychological present at the expense of cognitive futurity. *Journal of Consulting and Clinical Psychology*, 1972, *38*, 70–79.

Schilder, P. *Mind: Perception and thought in their constructive aspects.* New York: Columbia University Press, 1942.

Schlosberg, A. Time perspective in schizophrenia. *Psychiatric Quarterly,* 1969, *43,* 22–34.

Seeman, M. V. Time and schizophrenia. *Psychiatry,* 1976, *39,* 189–195.

Seligman, M. E. P. *Helplessness: On depression, development, and death.* San Francisco: Freeman, 1975.

Shapiro, D. *Neurotic styles.* New York: Basic Books, 1965.

Shepard, R. N. The analysis of proximities: Multidimensional scaling with an unknown distance function. *Psychometrika,* 1962, *27,* 125–139.

Siegman, A. W. The relationship between future time perspective, time estimation, and impulse control in a group of young offenders and in a control group. *Journal of Consulting Psychology,* 1961, *25,* 470–475.

Siegman, A. W. Anxiety, impulse control, intelligence, and the estimation of time. *Journal of Clinical Psychology,* 1962, *18,* 103–105.

Sorokin, P. *Sociocultural causality, space, and time.* Durham, N.C.: Duke University Press, 1943.

Spielberger, C. D. Theory and research on anxiety. *In* C. D. Spielberger (Ed.), *Anxiety and behavior.* New York: Academic Press, 1966.

Stein, K. B., Sarbin, T., & Kulik, J. A. Future time perspective: Its relation to the socialization process and the delinquent role. *Journal of Consulting and Clinical Psychology,* 1968, *32,* 257–264.

Straus, E. Disorders of personal time in depressive states. *Southern Medical Journal,* 1947, *40,* 254–258.

Straus, E. *Phenomenological psychology.* New York: Basic Books, 1966.

Thayer, S., Gorman, B. S., Wessman, A. E., Schmeidler, G., & Mannucci, E. The relationship between locus of control and temporal experience. *Journal of Genetic Psychology,* 1975, *110,* 275–279.

Torgerson, W. S. Multidimensional scaling. I: Theory and method. *Psychometrika,* 1952, *17,* 401–419.

Toulmin, S., & Goodfield, J. *The discovery of time.* New York: Harper and Row, 1965.

Tucker, L. R., & Messick, S. J. An individual differences model for multidimensional scaling. *Psychometrika,* 1963, *28,* 333–367.

Wallace, M. Future time perspective in schizophrenia. *Journal of Abnormal and Social Psychology,* 1956, *52,* 240–245.

Wallace, M., & Rabin, A. I. Temporal experience. *Psychological Bulletin,* 1960, *57,* 213–236.

Weber, A. O. Estimation of time. *Psychological Bulletin,* 1933, *30,* 233–252.

Wessman, A. E. Personality and the subjective experience of time. *Journal of Personality Assessment,* 1973, *37,* 103–114.

Wessman, A. E., Gorman, B. S., Schmeidler, G. R., Thayer, S., & Mannucci, E. Personality and linear representation of temporal location. *Journal of Consulting and Clinical Psychology,* 1974, *42,* 194.

Wessman, A. E., & Ricks, D. F. *Mood and personality.* New York: Holt, Rinehart and Winston, 1966.

Whitrow, G. J. *The natural philosophy of time.* London: Nelson, 1961.

Whyman, A. D., & Moos, R. H. Time perception and anxiety. *Perceptual and Motor Skills,* 1967, *24,* 567–570.

Wohlwill, J. F. Methodology and research strategy in the study of developmental change. *In* L. R. Goulet & P. S. Baltes (Eds.), *Life span developmental psychology.* New York: Academic Press, 1970.

8

Editors' Introduction

Arthur Schlesinger, Jr. takes a broad view of time and examines how the very notion of time has emerged and has been transformed over the course of Western history.

Schlesinger observes that while time was viewed in ancient and medieval times as basically static and immutable, the extraordinary advance of science and technology in recent centuries has profoundly altered our ideas about the pace and character of life. The mounting velocity of history is viewed as the primary shaping force in the development of modern consciousness, where life is perceived as motion, not order; the future is more salient than the past; and the universe is seen as not complete but in a process of creation.

Schlesinger demonstrates how the consequences of time acceleration have fundamentally altered many aspects of culture: philosophy, the natural and social sciences, art, literature, and education. The deep anguish of modern consciousness concerns the fate of human personality itself; for how can personal integrity, the sense of identity, stand up to the assault of the changed and multiplying pressures of a world that will not slow down? Modern man is often disoriented, unsure of his ideas, his institutions, his values, and his own beliefs concerning purpose and meaning. The crises of modernity are the source of the confused contemporary search for alternative values and meanings. The challenge of the future is to devise the means whereby a rampant technology can be brought under social control.

The Modern Consciousness and the Winged Chariot

Arthur M. Schlesinger, Jr.

We are living in a world and time of apparently incessant intellectual revolution, a permanent revolution in a profounder sense than Trotsky ever envisaged, rushing us on past landmark, guidepost, and certitude into a world and time beyond our control and imagining.

"Had we but world enough, and time . . ."—I recognize that Andrew Marvell had in mind the imminence of mortality and not the acceleration of history. But, in another sense, we too are cheated of the amplitudes of world and time, and at our backs, we too always hear time's winged chariot hurrying near. I would venture to suggest that a fundamental cause of our disquietude, a source at once of exhilaration and desolation, is the ever-quickening pace by which the world and time speed by us. Change is no doubt inseparable from human history. But never in human history has change been so organized and compulsive as in the last three centuries, and never have the forces of change converged on a single generation—and, one is tempted to add, a single country—as these forces converge upon the last quarter of the 20th century in the United States.

Arthur M. Schlesinger, Jr. • The Graduate School and University Center, City University of New York, New York. This paper is adapted from an address presented at the Tenth Anniversary Convocation of the Graduate School of the City University of New York, May 4, 1973.

1. The Acceleration of Modern Life

The cause of this change is the extraordinary and inexorable advance in science and technology. Of course, it can be said that science and technology have been advancing inexorably ever since the first caveman fashioned a club or lit a fire. But for most of our long travail on earth, the advance has been intermittent, limited, and halting. The change in phase began only about three centuries ago—a flashing second in the long span of human existence. William James pointed out in 1895 that there had been only four lifetimes since Galileo (James, 1897), and we have had only one lifetime since James. But each lifetime has seen stunning developments in scientific discovery, and the last two lifetimes have seen wondrously increasing proficiency and rapidity in the translation of science into technology—a development that, Whitehead (1925) has told us, distinguished the 19th century "from all its predecessors." Whitehead added, "The greatest invention of the nineteenth century was the invention of the method of invention" (chap. 6). The process of change had been through the long reach of history slow, unconscious, and unexpected. The 19th century made it suddenly quick, conscious, and expected. The new method, in Whitehead's (1925) words, broke up "the foundations of the old civilisation" (chap. 6).

Some years ago, the Bell Telephone Laboratories studied the time lag between scientific discovery and technological application. The principle of the electric motor, for example, was discovered in 1821 and applied in 1886—a lag of 65 years. The principle of radio broadcasting was discovered in 1887 and applied in 1922—a lag of 35 years. But radar had a lag of 5 years between discovery and application, the atomic bomb 7 years, the transistor 3 years, the solar battery 2 years (Ginzberg, 1964; Mensch, 1970). Nothing better symbolizes the acceleration in the rate of change in our own lifetime than the reflection that it would have been quite possible for someone who watched the Wright brothers fly for a few seconds in the air at Kitty Hawk in 1903 to have watched Apollo 11 land men on the moon in 1969.

Henry Adams (1919) was the first American to take systematic note of the increase in the velocity of history. Persuaded that social energy was a form of mechanical energy and thus subject to physical law, he construed history in terms of the second law of thermodynamics, the law of the dissipation of energy. In this light the future looked grim.

"The world did not [just] double or treble its movement between 1800 and 1900," he wrote in 1909, "but, measured by any standard known to science—by horse-power, calories, volts, mass in any shape—the tension and vibration and volume and so-called progression of society were fully a thousand times greater in 1900 than in 1800" (Adams, 1919, p. 303). He did not see how civilization could survive the exponential increase in the rate of change. In 1903, he had written, "My figures coincide in fixing 1950 as the year when the world must go smash" (Adams, 1938, II). Two years later, still brooding on the problem "how long the world will take, at its present acceleration, to break its neck" (Adams, 1938, II), he gave it less time, fixing on the decade 1930–1940.

Inspired by Willard Gibbs, Adams now tried to apply Gibbs's rule of phase to history in search of the mathematical equations of acceleration. He concluded that the velocity of history increased by "the old, familiar law of squares. . . . that is to say, that the average motion of one phase is the square of that which precedes it" (Adams, 1919, p. 291). Thus, the Religious Phase lasted about 90,000 years, the Mechanical Phase lasted about 300 years, and "the next or Electric Phase would have a life equal to $\sqrt{300}$, or about seventeen years and a half, when—that is, in 1917—it would pass into another or Ethereal Phase, which, for half a century, science has been promising, and which would last only $\sqrt{17.5}$, or about four years, and bring Thought to the limits of its possibilities in the year 1921" (Adams, 1919, p. 308). Adams's final prediction—oddly anticipatory in its phrasing of the title of the last desperate book by another literary man who for a season sought solutions in science, H. G. Wells's *Mind at the End of Its Tether*—was plainly premature, nor did Adams intend it literally. Indeed, his whole effort must be taken not as a serious scientific inquiry but as an elaborate and powerful literary metaphor. Still, the point beneath the metaphor was unassailable.

No one knows this better than those of us who live in New York City at the end of the third quarter of the 20th century. Rushed on by the cumulative momentum of science and technology, the tension and vibration and volume and so-called progression of society are incalculably greater today then they were when Adams despaired 60 years ago. Nothing defines our age more than the furious and relentless increase in the rate of change. Science makes, dissolves, rebuilds, and extends our environment every day. The planet has changed more in the last hundred years than in the thousand years preceding; and, whether or not acceleration comes by the law of squares, our world will change more between now and the end of the century than it has changed since Adams tried to apply the rule of phase to history. It is the mounting velocity of history, I would contend, that has been the primary shaping force in the development of the modern consciousness. Whitehead captured part of the point when he said in 1925, "In the past human life was lived in a bullock cart; in the future it will be lived in an aeroplane; and the change of speed amounts to a difference in quality." (chap. 6). If he had said that in the future human life would be lived in a devastating process of acceleration in which the airplane was a rudimentary and transient phase, he would have suggested the larger dimensions of the qualitative change.

Since the beginning of time, men and women had lived in a world that, if it changed at all, changed very slowly. In static societies, there was no great difference between the present and the past. The experience of older generations provided sufficient guidance for the young. The functional need for new ideas was limited. Society could subsist on the existing stock of wisdom for a long time. Tradition was sacred and controlling. But the shift from a slowly changing to a swiftly changing society has altered many things. It has defined the present as something radically different from the past. It has left in the distance the familiar reference points and rituals that had stabilized and sanctified life for generations. It has placed traditional institutions and roles under severe and incomprehensible strain. It has stretched and snapped customary social restraints. It has rendered the experience of the elders irrelevant to the problems of the young. Children, knowing they will live in a vastly different world from the world of their parents, no longer look to their parents as models and authorities. The acceleration of change has created an unending need for new knowledge and new explanation, so that even the young in the course of a lifetime must forget many things they have been carefully taught and learn new things in their place.

It has meant above all that life is perceived as motion, not as order; that the future is more of a presence than the past; that the universe is seen not as complete but in process of creation. For people of robust and buoyant temperament like William James the prospect was exhilarating. Others, like Adams, it filled with foreboding. Some, like Spengler and Toynbee and Yeats, foresaw the end of a civilization. And, as the velocity of history continued to intensify, it began to produce the widespread contemporary sense that the stimuli and pressures of acceleration are too much for either the human psyche or the social framework to absorb, that everything is getting out of control, that all circuits are breaking down from overload—till optimism has become the private property of B. F. Skinner and Mao Tse-tung and elsewhere a hundred flowers of pessimism bloom.

2. Change in Contemporary Philosophy and Science

So the modern consciousness has been shaped by the speedup in the rate of change. One of the primary distillates of consciousness is philosophy. In classic philosophy, metaphysics had been the study of absolute being. If change were admitted as part of life, it was change according to fixed and unchanging laws of being. The ancient philosophers, Bergson (1965) said, imagined "that being was given once and for all, complete and perfect," in an ultimate system of essences; the visible world unfolding "before our eyes could therefore add nothing to it; it was, on the contrary, only diminution or degradation." In such a world, time was an intruder, the "disturber of eternity" (pp. 55–56). As Dewey (1965) put it, classic philosophy saw change, and consequently time, as "marks of inferior reality" (pp. 208–209).

Such a view of reality was the natural projection of a slowly changing society. But the doctrine of reality as given from the start was hard to believe in a world commanded and convulsed by change. As a young man, Bergson had been much attached to the philosophy of Herbert Spencer until, as he put it, "I perceived one fine day that, in it, time served no purpose, did nothing" (Bergson, 1965, p. 46). To Bergson, Spencerism all of a sudden seemed nonsense; for what was reality but "the continuous creation of unforeseeable novelty" (Bergson, 1965, p. 44)? James (1911, chap. 9) similarly saw "the everlasting coming of concrete novelty into being" as the "obvious" truth of life. Both Bergson and James were determined to overthrow what Dewey later described as "the doctrine of the subordination of time and change" and to install "change at the very heart of things" (Dewey, 1965, p. 211). As Samuel Alexander put it in 1921, the distinguishing mark of modern philosophy was that it "takes time seriously" (Alexander, 1965, p. 163).

So philosophy, at least in one of its characteristic currents, shifted its concern from eternity to time, from immutability to change, from absolutes to processes and relations, from being to becoming. "No philosopher takes time seriously," wrote Whitehead (1965), "who *either* conceives of a complete totality of all existence, *or* conceives of a multiplicity of actual entities such that each of them is a complete fact" (p. 313). The modern consciousness, compelled by changes in the external environment to a vivid awareness of time, mutability, novelty, chance, and interconnection, saw the universe as open, indeterminate, and interrelated, with time, in Bergson's phrase, as the "vehicle of creation and of choice" (Bergson, 1965, p. 46). In the later works of Whitehead, process itself became reality; and though contemporary philosophy may have been more influenced for a season by the Whitehead of *Principia Mathematica* than by the Whitehead of *Process and Reality,* the modern consciousness has not sought to repeal the vision of James and Bergson and to reinstate the vanished realm of absolute essences.

The effort to place change in the center of philosophy both reflected and reinforced parallel developments in science itself. The 19th century had not only been the age of technology and of what Carlyle sarcastically called Victorious Analysis; it had also been the century in which science supposed it had established its intellectual foundations. This was the time—and here, as an inmate of the other culture, I must acknowledge my debt to Whitehead, Bronowski, and other missionaries to the infidel—when the method of science was "the manipulation of exact measurements" and when the philosophy implied by that method, the philosophy that took matter and mind as "independently existing substances," was the philosophy of scientific materialism and determinism. Lord Kelvin proclaimed the faith that scientists have long since renounced, though, alas, it has recently become the credo of a school of historians: "If you can measure that of which you speak, and can express it by a number, you know something of your subject. If you cannot measure it, your knowledge is meagre and unsatisfactory." "Science," in Bronowski's words, "came to be admired as the Puritan guardian of literal truth, whose aim it was to describe the material world visibly to the last decimal point" (Bronowski, 1970, pp. 3, 6; Whitehead, 1925, chap. 9).

Scientific materialism, so solid and invincible a century ago, has mostly faded away, except in intellectually underdeveloped countries like the Soviet Union. Scientific inquiry uncovered too many phenomena that determinist assumptions could not explain; and the scientific community, in the manner so cogently described by T. S. Kuhn (1962), proceeded to ring out the old paradigms and ring in the new. (See also Lakatos and Musgrave, 1970.) If some of the Kuhnian anomalies were the result of new scientific discoveries, some too were the result of a felt discordance between the old positivist science and the new perceptions of time, novelty, and context. Whitehead (1925), describing Wordsworth's reaction to Newtonian science, said, "He felt that something had been left out, and that what had been left out comprised everything that was most important." So, as we have seen, Bergson reacted to Spencer; and so Whitehead himself issued his call for an organic conception of the world that would "end the divorce of science from the affirmations of our aesthetic and ethical experience" (chaps. 5, 9).

Contemporary science is only preliminarily a matter of facts and measurement. It is essentially concerned, I take it, with the way facts are organized by underlying structures; and along the way it has conceded a great deal to ideas of indeterminacy and relativity. It has replaced absolute law by statistical probability, has decided that the experimenter may affect the experiment, and has acknowledged the existence of phenomena not readily explicable by conventional science, from extrasensory perception to acupuncture. Like philosophy, it has been suffused by the ever-intensifying velocity of history with an increasing awareness of the "black holes" in scientific understanding.

The acceleration of change has also left its mark on the social sciences, though here the impact has taken the form less of the installation of change at the very heart of things—for the social sciences had long been concerned with aspects of change—than of a complex and manifold enlargement of consciousness. For acceleration raised to the surface a whole range of issues that static societies never had to confront or articulate. When a society changes slowly, its presuppositions go for generations without examination. "Such assumptions appear so obvious," Whitehead wrote, "that people do not know what they are assuming because no other way of putting things has ever occurred to them" (Whitehead, 1925, chap. 3). But the age of acceleration was characterized by the obsolescence or collapse of old truths and the rise of new ones; it was characterized too by warfare between tradition and novelty; it involved a steadily more comprehensive criticism of the past by the future. People began to discover their assumptions when other ways of putting things became available and even attractive. And the faster the rate of change, the greater the need for new conceptions and the sharper the competition among diverging ideas.

People thus became newly aware of the axioms of their lives, newly aware of the dependence of "truth" on particular social and psychological situations. Knowledge became systematically self-critical as never before, and social thought acquired a new sensitivity to its own premises, self-interests, and illusions. This diverse enlargement of consciousness embraced a variety of fields and may be evoked by such phrases as the sociology of knowledge, psychoanalysis, neo-orthodox theology, cultural anthropology, analytic philosophy, institutional economics, intellectual and comparative history, symbolic forms, the New Criticism. To list the names of those involved is to call the roster of those who have formed the modern consciousness. What they provided, from Freud to Niebuhr, from Moore to Wittgenstein, from Pareto to Cassirer, from Jung to Weber, from Ortega y Gasset and Keynes to Lévi-Strauss and Lippmann and Galbraith, were a series of means by which consciousness could reflect more powerfully on itself.

3. Modern Art and Becoming

The social and psychological sciences hoped to drag the issues of life to the surface. But their instruments, though more elaborate than in earlier times, were still gross and blunt; and the deeper exploration of the unconscious remained as before the province of art. Writing, painting, sculpture, music—all were swept into the rapids of change. Here too, the modern consciousness had its very distinctive timbre. Ortega y Gasset noted in *The Revolt of the Masses* (1932) how various periods of history have cherished the illusion of "the plenitude of time"—the illusion "that the end of a journey has been reached, a long-felt desire obtained, a hope completely fulfilled." The 19th century had been preeminently one of those "self-satisfied" ages. The 20th century could no longer regard itself as definitive; on the contrary, it was sure there were no such epochs, final "assured, crystalized forever" (chap. 3). In the modern consciousness, the present had its value not as the culmination of the past but as the prelude to the future. In art as in philosophy, becoming had superseded being.

"The lust for otherness, for newness," wrote Bernard Berenson in 1948, "which seems the most natural and matter-of-course thing in the world, is neither ancient nor universal. Prehistoric races are credited with having had so little of it that a change in artifacts is assumed to be a change in populations" (Berenson, 1954, p. 155). This special tone of the modern consciousness can be suggested by contrasting Emerson's criticism of the early 19th century—"Our age is retrospective. It builds the sepulchres of the fathers"—with Mayakovsky's exultant cry: "Make bombardment echo on the museum walls. . . . Why didn't they string up Pushkin?" (Poggioli, 1971, p. 53).

Of course, this is not all the modern consciousness has had to say about the past. It has also understood the secret strength imparted by tradition. But the tradition of the modern was not "tradition" as embalmed in the academy. Rather, in the spirit of that great premodern Emerson, the modern artist has ransacked the past to invent his own tradition, often finding particular sustenance and stimulus in art the academy had forgotten. "Only an inventor," as Emerson said, "knows how to borrow" (Emerson, 1903a).

The acceleration of change thus denied the modern age the consolation enjoyed in other ages that it had attained to the fullness of time; instead, it saw itself as a time of anticipation and transition. And the law of acceleration deprived it of another consolation. A slowly changing world could evolve and preserve a relatively harmonious and abiding perspective on life. The aesthetic ideal of the West from the ancient Greeks through the 19th century had been as Matthew Arnold said of Sophocles, to see life steadily and see it whole. That possibility was another casualty of the increasing velocity of history.

Seeing life at once steadily and whole was plausible enough when the artist and his environment had a stable relationship over time. But when the artist found himself traveling, so to speak, at a thousand miles an hour across a landscape constantly taking new forms and colors, when time changed from an unexamined assumption into an interfering force, it was no longer possible both to see life steadily and to see it whole. Gertrude Stein, the sybilline prophet of the modern consciousness, well described the 20th century as "a time when everything cracks, where everything is destroyed, everything isolates itself," though, with all this, "a more splendid thing than a period where everything follows itself" (Stein, 1973). Artists in the 20th century, as Forster somewhere has it, had to make the choice: they could see life steadily, or they could see it whole, but they could not hope to do both at once.

If they sought to see life steadily, they had to narrow the scope and focus of their art—to renounce the moving landscape and concentrate on a single point. For a season, painting and poetry became microscopic, hoping to defeat time and evoke totality through minute representation of an impression or image: thus impressionist painting and imagist poetry—correctly associated by T. E. Hulme, himself a student of Bergson, when he lectured on poetry to the School of Images: "What has found expression in painting as Impressionism will soon find expression in poetry as free verse" (Hulme, 1963, pp. 14, 24). Ezra Pound well defined the joint enterprise when he wrote, "An 'Image' is that which presents an intellectual and emotional complex in an instant of time. . . . It is the presentation of such a 'complex' instantaneously which gives the sense of sudden liberation; that sense of freedom from time limits and space limits; that sense of sudden growth, which we experience in the presence of the greatest works of art" (Pound, 1963, p. 18).

But Impressionism and Imagism would not suffice for those who sought to see life whole. To achieve wholeness, art had to abandon steadiness. It had to devise techniques of dislocation, discontinuity, and penetration to express the disorientation of an age wracked by time. Gleizes and Metzinger (1964) in their essay on Cubism condemned Courbet and the Impressionists equally for permitting the eye to predominate over the brain: "Unaware that in order to discover one true relationship it is necessary to sacrifice a thousand surface appearances, [they] accepted without the slightest intellectual control everything [their] retina communicated." "Imitation," in the Cubist view, "is the only error possible in art; it attacks the law of time, which is Law" (p. 3).

Cubism was a first try at integrating the plastic consciousness and discovering the dynamism of form in time. The search for modes by which the quality of modern life could be expressed involved restless experimentation. This is notable in the three artists who carried the logic of the modern consciousness to its extreme limits without betraying the deeper disciplines of the artistic tradition—Picasso, Joyce, and Stravinsky. It is equally clear in artists whose experiments were more restrained but who understood, as Bergson had, that time was now the vehicle of creation and of choice—Proust, Yeats, Eliot. *À la Recherche du Temps Perdu,* the century's greatest novel, was above all an exploration in time. The poetry of Yeats was saturated with an awareness of time. Eliot was perhaps the most conservative of all by temperament, most explicitly respectful of tradition. Yet he prefaced his greatest poem with an epigraph from Heraclitus, the one ancient for whom change was the essential quality of experience, and the first lines of the poem were

> *Time present and time past*
> *Are both perhaps present in time future,*
> *And time future contained in time past.*
>   (Eliot, 1944, pp. 6–7)

If Eliot sought a "still point" in *Four Quartets,* he was well aware that it was a "turning world." Nor, with a writer of Eliot's cunning, is it safe to suppose that he meant only one thing when he ended the second part of *The Waste Land* with the reiterated refrain "HURRY UP PLEASE IT'S TIME."

The search for techniques expressive of the disorder wrought by the acceleration of history characterizes modern painting, music, sculpture, and literature. It is recalled by such words as *symbolism* and *cubism, dadaism* and *surrealism, abstract expressionism* and *atonalism, polyphonic prose* and *twelve-tone music, the stream of consciousness* and *the interior monologue.* Since science and technology had made time the great protagonist in the modern consciousness, it was not surprising that science and technology offered suggestions to art as they had to philosophy and the social sciences. The experimental method itself was practiced as devotedly by Picasso, Joyce, and Stravinsky as by any scientist in his laboratory. But it went much further than this. Science provided new ways of looking at the world—ways that reinforced the artist's quest for techniques that would catch and convey the discordant and disconnected vibrations of the age.

Once again Emerson had foreseen the future. "The correlation of forces and the polarization of light," he wrote in 1867, "have carried us to sublime generalizations,—have affected an imaginative race like poetic inspirations" (Emerson, 1903b). "A scientific event cleared my way of one of my greatest impediments," Kandinsky wrote half a century later. "This was the further division of the atom. The crumbling of the atom was to my soul like the crumbling of the whole world" (Kandinsky, 1964, p. 27). As realism and naturalism in literature and painting had expressed the positivist presuppositions of 19th-century science, so 20th-century art paralleled 20th-century science in the shift of concern from exact measurement and surface representation to the quest for inner structures and general relationships. Mondrian argued for nonfigurative art in 1937 because it showed that "'art' is *not the expression of the appearance of reality . . . but . . . the expression of true reality*." Representational art could aspire at best to particular and visible reality; abstract art aimed at objective reality. "While universal reality," Mondrian wrote, "arises from determinate relations, particular reality shows only veiled relations" (Mondrian, 1964, p. 127). And what after all was Yeats's "How can we know the dancer from the dance" (Yeats, 1950, p. 245) but a poetic rendition of the proposition of 20th-century science that scientific observation cannot be separated from the observer? or Forster's "Only connect . . . live in fragments no longer" (Forster, 1910, chap. 22) but a command to science as well as to life?

"We study in crystals not what they are made of," Bronowski wrote, "but how they are put together." The artistic pursuit of structures that lay deeper than the skeleton seemed to him to express the new scientific vision. Thus, Bronowski found in the sculpture of Henry Moore a realization of the topological relations so central to the outlook of contemporary science. He concluded that, while critics had argued which culture had the monopoly of the new vision, "artists and scientists have gone quietly about their business of feeling and expressing the same common revolution" (Bronowski, 1970, pp. 9–11).

Technology as well as pure science affected the artistic vision. It was not only the invention of new media of expression—film, radio, television, electronic music—though this was of the highest importance; it was also the extent to which technological change brought the physical world itself into new perspectives. Here too, artistic imagination paralleled or anticipated technological innovation. Let us remember Whitehead's suggestion about the qualitative transformation induced by the change from the bullock cart to the airplane. Gertrude Stein, traveling across America by plane in the thirties, looked at the earth falling away below and was astonished to see "all the lines of cubism . . . the mingling lines of Picasso, coming and going, developing and destroying themselves . . . the simple solutions of Braque . . . the wandering lines of Masson." Yet the painters had thus seen the world 30 years before, "when not any painter had ever gone up in an airplane." But Picasso, for example, being of the 20th century, "inevitably knew that the earth is not the same as in the nineteenth century, he knew it, he made it, inevitably he made it different and what he made is a thing that now all the world can see. . . . He understands what is contemporary when the contemporaries do not yet know it" (Stein, 1973). So the artist in the age of acceleration derives his vitality from his prevision of what is to come.

4. The Challenge to Personal Identity

We have seen how the mounting velocity of history has infused the modern consciousness with a new and passionate awareness of time, mutability, novelty, and chance and how that consciousness has therefore sought new modes of understanding and creation in philosophy, including the philosophy of science, in the social sciences, and in the arts. But these have been strategies of intellectual survival by which the creative personality has tried to keep abreast of the world where, in the old phrase of Aristophanes, whirl is king. The deeper anguish of the modern consciousness is the fate of the underlying human personality itself. How can the integrity of personality, the sense of identity, stand up to the assault of change, to the constant multiplication of pressures and of stimuli, to the infernal bombardment leveled against us all by a world that will not slow down?

For personality found its easiest integration in relatively static societies, where roles, expectations, values, and identities were formed by immemorial traditions and controlled by unchanging structures. The world of change has liberated men and women from such traditions and structures. At the same time, to a greater degree than ever before in history, it has sent individuals out on their own to construct from their own resources their roles, expectations, values, and identities. For the newly liberated generation, there was a moment of exultant release—thus the Emersonian hymns to self-reliance. But even Emerson's perceptions were tinged with a melancholy sense of the "evanescence and lubricity" (Emerson, 1906) of experience, and some of his contemporaries saw with savage clarity the dark underside of the new freedom. Melville recorded his dissent in that most desperate novel *Pierre* with its significant subtitle *Or, the Ambiguities.* By the early 20th century, Yeats met the problem of identity by evolving the doctrine of masks—an idea that in the centuries preceding had served only as a component of farce. As he wrote in 1909, "I think that all happiness depends on the energy to assume the mask of some other self; that all joyous or creative life is a re-birth as something not oneself, something which has no memory and is created in a moment and perpetually renewed" (Ellman, 1958, p. 174).

In the years since 1909, the rate of change has gathered momentum. It is little wonder that, as we all flounder in the contemporary maelstrom, we feel we are drowning in the ambiguities: forever disoriented and off balance; unsure of our ideas, institutions, and values; unsure of our relations to others and to society; unsure of our own purpose and identity. As the world has accelerated, it has become increasingly plain that, if the psychic cost of slowly changing societies was the stultification of personality, the psychic cost of swiftly changing societies is the disintegration of personality.

Again the social literature is filled with description—Lifton's protean man, Riesman's other-directed man, Whyte's organization man, not to mention Eliot's hollow men, Bellow's Herzog, and Hesse's Steppenwolf. All portray modern man as "Shape without form, shade without colour,/Paralysed force, gesture without motion" (Eliot, 1963), forming prayers to broken stone, existing not through an inner core of being but in response to external cues and cries: man, in short, as tropism. Sartre thus defined consciousness as "a great emptiness, a wind blowing toward objects" (Lifton, 1968).

The modern consciousness is thus at once vague and vulnerable. Lacking the internal conviction of identity instilled by traditional society and sanctified by traditional religion, it launched itself on long and fruitless quests for deliverance and unity. When the rate of social change is slow, as Professor Trilling has so finely demonstrated, the goal is sincerity, the discovery of self; when the rate is rapid, the goal becomes authenticity, the manufacture and verification of self.

5. Counterrevolutions and the Search for New Ideologies

In its most defiant mood, the modern consciousness has starkly accepted the death of God and the absurdity of life and called on man to create his own identity through his confrontations and his choices. But the existential view of life, I cannot help but think, required preexisting strength and purpose; it required men like Russell, Camus, and Sartre;

in the end, it presupposed what it purported to create. Others in their quest for authenticity donned and doffed masks, tried out and discarded beliefs, became spiritual window-shoppers (Pegler's phrase about Henry Wallace), ran through one nostrum after another with deep conviction and at top speed, last year encounter groups and Zen Buddhism, this year astrology and organic foods, next year—who knows?—hydropathy or Zoroastrianism or Guru Maharaj Ji. The frantic search for true consciousness ended in a mess of false consciousnesses.

The trouble when people stopped believing in God, as Chesterton so well said, was not that they thereafter believed in nothing; it was that they thereafter believed in anything. The strain of freedom at best could be unbearable, as Dostoevski showed in the fable of the Grand Inquisitor; and, in a world forever destroyed and reassembled by the mounting velocity of history, people began to flee from freedom. Thirty years ago Erich Fromm explained how the "escape from freedom" opened Europe to totalitarianism. Certainly, a potent appeal of fascism and of communism was the presumed replacement of loneliness by comradeship, of isolation by solidarity. In our own time, the escape from freedom has taken a variety of forms, all promising the restoration and reintegration of personality. Frantz Fanon tells us that the individual, at least in colonial countries, will find himself through acts of violence. Others seek salvation by withdrawal from the accelerating world into the supposed simplification and stability of communal living. Others try to transcend the fragmentation of consciousness by drugs.

The line between illusion and reality has grown thinner through the century. Pirandello and Pinter have disclosed how enigmatic fantasy pervades commonplace life. Lately, we have seen the theory arising that madness is the form sanity takes in an insane world—a proposition first advanced with some felicity against the background of world war in such novels at *The Good Soldier Schweik* and *Catch 22* and subsequently reduced to solemn sophistry by writers like R. D. Laing.

Behind this all gathers a crusade against reason itself and against science as the most evil and effective form of reason. Science in a sense has invited this reaction not only by putting the world in so insensate a whirl but by its philosophical pretensions along the way, by its complacent denial of phenomena it could not explain, by the instruments of destruction it has placed in the service of power, by its ravages of the natural environment, by its evident determination to meddle with the mysteries of personality through such means as genetic engineering and psychosurgery. One applauded when James D. Watson, who shared the Nobel Prize for cracking the genetic code, called for an end to experimentation with human cell fusion and embryos, saying, "If we do not think about the matter now, the possibility of our having a free choice will one day suddenly be gone" (Watson, 1971).

Jacques Monod too may well be right in seeing even deeper reasons for the new warfare against science—"the fear of sacrilege: of outrage to values." The modern consciousness, he has suggested, is rent by an agonizing contradiction between the old values inherited from immemorial animistic traditions—"a disgusting farrago of Judeo-Christian religiosity, scientistic progressism, belief in the 'natural' rights of man, and utilitarian pragmatism"—and the new values required by the search for objective knowledge. Science, in short, has conquered its place in men's practices but not in their hearts. Scientific practice had launched the evolution of culture on a one-way path: "onto a track which nineteenth-century scientism saw leading infallibly upward to an empyrean noon hour for mankind, where what we see opening before us today is an abyss of darkness." Monod, and Bronowski too, argued that science itself, properly construed, becomes a source of value and that "the ethic of knowledge" can cure the schism in the modern soul (Monod, 1972, pp. 170–177).

In any case, the crusade—or should it be called the pogrom?—has begun against the scientific style of mind and against reason itself. "Reason," Theodore Roszak said recently in a splendid outburst of reductionism, ". . . is the god-word of a specific and highly impassioned ideology handed down to us from our ancestors of the Enlightenment as part of a total cultural and political program. Tied to that ideology is an aggressive dedication to the urban industrialization of the world and to the scientist's universe as the only sane reality." Against the reign of science and reason the counterculture calls for immersion in the unconscious, for the cultivation of the faculties of meditation and mysticism, for experiencing "the irrational discriminately and generously for what it is—a rich spectrum of life-enhancing human possibilities" (Roszak, 1973). Forster perhaps put it better some 60 years ago in *Howards End* when Margaret Schlegel, noting that "one's hope was in the weakness of logic," says, "This craze for motion has only set in during the last hundred years. It may be followed by a civilization that won't be a movement, because it will rest on the earth" (Forster, 1910).

There is great value in this as a warning to science and, I believe, something of value in this for itself—even though one sometimes feels that Asia has taken its revenge for our attempt to westernize the East by arranging for the orientalization of the West. The counterrevolution against science can easily go too far. Do many people really wish the decline in living standards and the increase in social tensions that would accompany the cessation of growth? Even the counterculture relies heavily on its motorcycles, its electric guitars, its tape recorders, and its hallucinogens; and few of these things grow on bushes. One particularly regrets the vulgarity of the attack on the exploration of space. I for one continue to think that the landing of men on the moon is the most exciting event of our age and that it will be remembered long after everything else about the 20th century is forgotten.

6. Today and Tomorrow

But the counterrevolution against science, like the reign of science, itself, is a function of the velocity of history. Nor can we suppose that the rate of change is likely to slow down. The "law of acceleration," Henry Adams wrote, "definite and constant as any law of mechanics, cannot be supposed to relax its energy to suit the convenience of man" (Adams, 1918, chap. 34). Adams's even more brilliant contemporary, William James, observed that humanity had thus far experienced only the slightest impact of science and technology. "Think how many absolutely new scientific conceptions have arisen in our own generation," he wrote in 1895, "how many new problems have been formulated that were never thought of before, and then cast an eye upon the brevity of science's career. . . . Is it credible that such a mushroom knowledge, such a growth overnight as this, *can* represent more than the minutest glimpse of what the universe will really prove to be when adequately understood? No! Our science is a drop, our ignorance a sea" (James, 1897, pp. 53–54). It is chastening to consider the fact, so often remarked upon by our scientific brethren, that, of all the scientists who have ever lived in the history of the planet, 90% are alive today—and all, it must be added, hard at work increasing the processes of change.

These processes are already rushing us on from the mechanical age to the Electric Age, if at a somewhat later date than Adams had predicted: from the age created by the steam engine, the railroad, and the assembly line into the fantastic new age of electronic informational and communications systems and electronic mechanisms of forecast, feedback, and control, the epoch foreshadowed by television, the computer, the laser, and the space satellite. Marx told us that history was determined by changes in the means of production, and there was something to that. McLuhan now tells us that history is determined by changes in the means of communication, and there is something to that too. One does not have to be a devout McLuhanite to recognize that underneath the extravagance and the vaudeville there is a point. Movement into the electronic age will have penetrating effects not alone on the processes of social organization but on the very reflexes of individual perception.

Certainly this movement has already introduced another complexity into the modern consciousness. Where the culture of print gave experience a frame and viewed it in sequence and from a distance, electronic communication—instantaneous, simultaneous, and collective—promises to undermine linear ways of thought. The print culture programmed the mind in a one-at-a-time, step-by-step way; its distinctive qualities were logic, precision, specialization, individualism. The electronic culture is not one-at-a-time but all-at-once; the distinct and separate are under challenge by the flowing, the unified, the fused.

All the pressures of change have risen in intensity and force in the last generation in the United States and have placed increasing strain on inherited ideas, institutions, and values. The old ways by which we reared, educated, employed, and governed our people seem to make less and less sense. The result of the injection into society of unprecedented and uncontrolled change has been to burst the traditional framework of value and institution in which we lived. Far from making that framework more cohesive and effective than ever before, as Professor Marcuse oddly argues, the onward roar of change has begun to fragment it on every side. Hence we have what is termed the "permissive society." So far as speech is concerned, whether political or pornographic, I don't suppose it has ever been more unrestrained in America than it is today. So far as people's appearance, dress, behavior, or even sense of their own sexuality are concerned, I don't think there has ever been a time in this country when individual variations have been more pronounced or choice more unfettered. The very invention of that dubious word *lifestyle* is an acknowledgment of the diversity of ways in which people have come to live. And, while permissiveness has produced its critics, the self-appointed keepers of our national morals turn out to have been the most permissive of all, indulging in a disregard for law and decency so squalid that it would make many a hippie spit with contempt. Still, permissiveness, by multiplying choice, may also multiply anxiety and fuel the escape from freedom.

The crisis of the modern consciousness, harried by change and by time, seems today more acute in the United States than anywhere else. The unwary or the ideological have claimed this as the result of the innate depravity of our economic system or of our national character. It would be more just, I believe, to recognize simply that the world is changing faster today than it ever has before and has been changing fastest of all in the United States. The revolutions wrought by science and technology have gone furthest here. As the nation at the extreme frontier of technological development, America has been the first to experience the unremitting trauma of accelerated change. The crises we are living through are the crises of modernity. Every nation, as it begins to reach a comparable stage of technical development, will undergo comparable crises—whatever their system of ideology or ownership. If anyone supposes that communism, for example, will solve the ecological crisis, let him swim in the waters of Lake Baikal.

The experience of accelerated change is already creating a common consciousness that to a degree is undermining the warfare of creeds. The Russians and soon the Chinese are discovering that the velocity of history spins off problems for which Marx and Lenin had no answer; which cannot even be solved by the little red books of the Chairman. Indeed, the totalitarian system is more vulnerable than the democratic system to the winged chariot, because the more centralized the control and the more absolute the ideology, the more rigid and brittle the system itself becomes, the less capable, so to speak, of rolling with the punch of history.

What this all means for higher education, I cannot say—and luckily have little enough time left to display the poverty of my thought. I can only make the obvious suggestion—heard so often in my lifetime, acted upon so rarely—that the history of science and technology ought to be a vital part of every academic curriculum. For, whatever claim to primacy humane knowledge in the classical sense may have had during most of the life of man on this planet, that primacy has come to an end in the age of acceleration. The West, so long dominated by the written

word and sequential logic, is now haunted by the rise of "languages outside the word"—I believe the phrase is George Steiner's—especially
mathematical and symbolic notation. "As electronic data-processing and
coding pervade more and more of the economic and social order of our
lives," Steiner warns us, "the mathematical illiterate will find himself
cut off. A new hierarchy of menial service and stunted opportunity may
develop among those whose resources continue to be purely verbal"
(Steiner, 1971).

And, with all this, we need to understand that science and
technology are in the true sense humanities too. "It is essential to keep
in mind," wrote Whitehead in 1921, "that science and poetry have the
same root in human nature." Education, he continued, must give us
"the feeling for the important sort of scientific ideas and for the
important ways of scientific analysis" (Whitehead, 1965, pp. 46, 49). We
need this to enrich our insight into the mysteries of the human mind and
sensibility. And we need it too, of course, in order to grasp the forces
that shape and dominate the world in which we live. We need to understand the past of science and technology—the intellectual impulses, the
social needs and perils—if we are to have a chance to meet the supreme
problem of the future, which is, I believe, to devise the means by which
a rampant technology may be brought under social control.

The modern consciousness is, in the language of the racing form,
by time out of memory. It is, by definition, in acute flux and schism. It
is the consciousness of uncontrolled change and therefore of incipient
chaos. The center, the greatest poet of the century tells us, cannot hold,
and the rough beast slouches toward Bethlehem to be born. It is hard in
these matters to decide between the prosaic and the prophetic, between
the premonition that things will go on much as they are and the premonition of catastrophe. Yet one supposes an ultimate tenacity in human-
kind, an abiding vitality in the human mind as well as in the human
spirit. So consider Andrew Marvell:

> Let us roll all our strength and all
> Our sweetness up into one ball,
> And tear our pleasures with rough strife
> Through the iron gates of life:
> Thus, though we cannot make our sun
> Stand still, yet we will make him run.
>
>   (Marvell, 1652)

Consider this, and consider too MacLeish's 20th-century rejoinder:

And here face down beneath the sun
And here upon earth's noonward height
To feel the always coming on
The always rising of the night . . .
To feel how swift how secretly
The shadow of the night comes on. . . .

(MacLeish, 1952, pp. 50–51)

In seasons of doubt and dispirit, I return to—and here I conclude with—the wisest of Americans, who said in his address on "The American Scholar" on August 31, 1837: "If there is any period one would desire to be born in—is it not the era of revolution when the old and the new stand side by side and admit of being compared; when all the energies of man are searched by fear and hope; when the historic glories of the old can be compensated by the rich possibilities of the new era? This time like all times is a very good one if one but knows what to do with it" (Emerson, 1901).

References

Adams, H. *The education of Henry Adams.* Boston: Massachusetts Historical Society, 1918.

Adams, H. The rule of phase applied to history. *The degradation of the democratic dogma.* New York: Macmillan, 1919.

Adams, H. Letters of January 25, 1903, and May 3, 1905. *In* W. C. Ford (Ed.), *Letters.* Boston: Houghton-Mifflin, 1938.

Alexander, S. *In* D. Browning (Ed.), *Philosophers of process.* New York: Random House, 1965.

Berenson, B. *Aesthetics and history.* New York: Anchor Books, 1954.

Bergson, H. The possible and the real. *In* D. Browning (Ed.), *Philosophers of process.* New York: Random House, 1965.

Bronowski, J. The discovery of form. *In* E. M. Jennings (Ed.), *Science and literature.* New York: Anchor Books, 1970.

Dewey, J. Time and individuality. *In* D. Browning (Ed.), *Philosophers of process.* New York: Random House, 1965.

Eliot, T. S. *Four quartets.* London: Faber & Faber, 1944.

Eliot, T. S. The hollow men. *In Collected poems, 1909–1962.* New York: Harcourt, Brace Jovanovich, 1963.

Ellman, R. *Yeats: the man and the masks.* New York: Dutton, 1958.

Emerson, R. W. The American scholar. New York: Laurentian Press, 1901. (Originally published, 1837.)

Emerson, R. W. Quotation and originality. *Letters and social aims.* Boston: Houghton Mifflin, 1903a. (Originally published, 1867.)

Emerson, R. W. Progress of culture. *Letters and social aims.* Boston: Houghton Mifflin, 1903b. (Originally published, 1867.)

Emerson, R. W. Experience. *Essays.* New York: E. P. Dutton, 1906. (Originally published, 1844.)

Forster, E. M. *Howards end.* London: G. P. Putnam, 1910.

Ginzberg, E. (Ed.) *Technology and social change.* New York: Columbia University Press, 1964.

Gleizes, A., & Metzinger, J. Cubism. *In* R. L. Herbert (Ed.), *Modern artists on art.* Englewood Cliffs: Prentice-Hall, 1964.

Hulme, T. E. Quoted in introduction, in W. Pratt (Ed.), *The imagist poem.* New York: Dutton, 1963.

James, W. *The will to believe.* New York: Longmans, Green & Co., 1897.

James, W. *Some problems of philosophy.* New York: Longmans, Green & Co., 1911.

Kandinsky, W. Reminiscences. *In* R. L. Herbert (Ed.), *Modern artists on art.* Englewood Cliffs: Prentice-Hall, 1964.

Kuhn, T. S. *The structure of scientific revolutions.* Chicago: University of Chicago Press, 1962.

Lakatos, I., & Musgrave, A. *Criticism and the growth of knowledge.* Cambridge: Cambridge University Press, 1970.

Lifton, R. J. Quoted in D. R. Cutler (Ed.), *The religious situation.* Boston: Beacon, 1968.

MacLeish, A. You Andrew Marvell. *Collected poems.* Boston: Houghton Mifflin, 1952.

Marvell, A. To his coy mistress. 1652. *In* A. M. J. Smith (Ed.), *Seven Centuries of Verse.* New York: Scribners, 1957.

Mensch, G. The average speed of social change. . . . paper, April 1970.

Mondrian, P. Plastic art & pure art. *In* R. L. Herbert (Ed.), *Modern artists on art.* Englewood Cliffs: Prentice-Hall, 1964.

Monod, J. *Chance and necessity.* New York: Vintage, 1972.

Ortega y Gasset, J. *The revolt of the masses.* New York: W. W. Norton, 1932.

Poggioli, R. *The theory of the avant-garde.* New York: Icon, 1971.

Pound, E. Quoted in W. Pratt (Ed.), *The imagist poem.* New York: Dutton, 1963.

Roszak, T. Some thoughts on the other side of this life. *New York Times,* April 12, 1973.

Stein, G. On Picasso in 1938. *New Republic,* April 21, 1973.

Steiner, G. A future literacy. *Atlantic Monthly,* August, 1971.

Watson, J. Quoted in The Washington Post, October 13, 1971.

Whitehead, A. N. The place of science in general education. *A philosopher looks at science.* New York: Philosophical Library, 1965.

Whitehead, A. N. *Science and the modern world.* New York: Macmillan, 1925.

Whitehead, A. N. Time. *In* D. Browning (Ed.), *Philosophers of process.* New York: Random House, 1965.

Yeats, W. B. Among school children. *Collected poems.* London: Macmillan, 1950.

Index

289